Working in Engineering

A guide to qualifying and starting a successful career in engineering

Professor Tony Price

with contributions from Dr Sophie Reissner-Roubicek

Working in Engineering

This first edition published in 2014 by Trotman Education, an imprint of Crimson Publishing Ltd, The Tramshed, Walcot Street, Bath BA1 5BB.

© Trotman Education 2014

Author: Tony Price

The right of Tony Price to be identified as the author of this work has been asserted by him in accordance with the Copyright, Designs and Patents Act 1988.

British Library Cataloguing in Publication Data
A catalogue record for this book is available from the British Library

ISBN 978 1 90604 199 1

Typeset by IDSUK (DataConnection) Ltd
Printed and bound in the UK by Bell & Bain Ltd, Glasgow

Contents

Contents

Contents

About the authors

Professor Tony Price is a Chartered Engineer and a Member of the Institute of Materials, Minerals and Mining and has over 30 years of experience in higher education and engineering. He has held oversight for the taught degrees within the School of Engineering at the University of Warwick since 1997. The School of Engineering adopts a unified approach to engineering and offers a range of engineering degrees that grow out of a common two years of study. He has been instrumental in the development of the curriculum and was responsible for working with the British Tunnelling Society to introduce an MSc in Tunnelling and Underground Space to help address the industry's shortage of qualified personnel. He teaches Rock Mechanics on this MSc and Geotechnical Engineering to civil engineering undergraduates.

Dr Sophie Reissner-Roubicek is a Senior Teaching Fellow at the Centre for Applied Linguistics at the University of Warwick. Prior to this she was based at the University of Auckland where her doctoral study of job interviews was conducted. A key part of the data analysed to confirm and describe what linguistic and interactional patterns or problems typically occur in novices' behavioural, or competency-based, job interviews was collected from engineering students and engineering company recruiters. Thanks to the insights gained from studying how engineers communicate, her research areas now extend to teamwork and leadership in intercultural contexts, and in her current role she develops and delivers intercultural workplace training courses for postgraduate engineers at the University of Warwick.

Acknowledgements

I am indebted to many colleagues, students, graduates and friends who have given their advice and support to the writing of this book. Any errors in this book are entirely mine and should not be attributed to anybody listed here. I am especially grateful to the following colleagues:

- Dr Sophie Reissner-Roubicek, Centre for Applied Linguistics: for her contribution on job interviews for graduate positions in engineering, including outcomes for female engineers and the pivotal role of teamwork questions

- Dr David Dyer, School of Engineering: for his advice on the content of the chapter on electrical and electronic engineering

- Ben Mayo, Royal Academy of Engineering Visiting Professor: for his advice on the content of the chapter on chemical engineering

- Stephen Ward, Head of External Relations and Siobhan Qadir, Senior Careers Consultant, both members of the University of Warwick's Student Careers & Skills department.

The content of this book is greatly enhanced by the *My story …* contributions. The contributors have been selected from individuals at different stages in their careers and are graduates from a range of universities. They are:

- Alasdair Woodbridge, self-employed, business start-up Heat Genius Ltd

- Dick Elsy, Chief Executive of the High Value Manufacturing Catapult

- Divya Surana, Business Analyst at Procter & Gamble and Brand Manager at Fizzy Foodlabs

- Ed Hall, Electrical Engineer at Siemens Wind Power

- Gaston Chee, self-employed, co-founder of BeGo

Acknowledgements

- Geoff Mayes, Project Manager at ITCM

- Kate Cooksey, Senior Tunnel Design Engineer at Morgan Sindall Underground Professional Services

- Leon Hawley, Assistant Engineer at URS Corporation

- Mukunth Kovaichelvan, Materials and Process Modeller at Rolls-Royce

- Navroop Singh Matharu, research student at University of Warwick

- Sarah Chen, mature student at University of Warwick

- Steve Dobson, General Manager for Product Support at JCB China.

The staff at Crimson have been very supportive of this project and I am indebted to Jessica Spencer, Holly Ivins and Libby Walden for their encouragement, constructive criticism and especially their patience.

Finally, I must thank my wife Pam for her support and patience during the many hours that I have committed to researching and writing this book. I would like to dedicate this book to her, Sarah, Jess, Emily and Adam.

Tony Price, 2013

Introduction

I f you are considering a career in engineering you are about to open up a world of opportunities for yourself. Moreover, you do so at a time when the importance of engineering and manufacturing is being reasserted by government and there are many new initiatives to build on the engineering strengths of the UK. At the same time the projected demand for qualified engineers is outstripping the supply. All this adds up to what should be a buoyant market for graduate engineers in the foreseeable future.

This book is written for those considering a career in engineering and will be of use for people of age 14 years upwards, including those considering a career change or sideways shift into engineering. It will be of particular use to those considering reading for a degree in engineering and offers useful advice on the skills that they should develop during their course and the importance of industrial experience. Engineering offers a diverse range of roles and opportunities for those with ability in science and mathematics, including opportunities to travel and work internationally. Engineering roles take several forms including technician, incorporated and chartered engineer and these are described briefly in Chapter 2; however, this book focuses on the education and employment opportunities for those seeking to become chartered engineers. For those interested in the other grades this book will provide both useful background and links to resources for the different disciplines.

This book is structured in three sections. Part 1 describes the motivations for becoming an engineer, the routes into the profession and the skills that an engineer develops. It also sets out the professional, statutory and regulatory bodies that monitor standards and promote the interests of engineering both nationally and worldwide. Part 2 focuses on the common engineering disciplines, providing information on the sectors in which engineers of that discipline find employment and on typical salary levels. Part 3 deals with the opportunities for and value of work experience; guidance on how to win your first graduate job; and a guide to what you might expect in terms of career development. It also discusses opportunities for postgraduate study or careers outside of engineering. At the back of the book you will find a glossary of terms that may be unfamiliar to you.

Introduction

In reading this book you will soon realise that the notional boundaries between different branches of engineering are remarkably porous, with each discipline bleeding into the next, revealing many overlaps of interest. Equally significant sectors of employment such as aerospace, automotive, building services and utilities will recruit graduates from a variety of disciplines because they need the breadth of skills that multidisciplinary teams offer. This is reflected throughout the book as chapters cross-reference each other to provide information on common sectors.

The book also contains a range of case studies in the sections called *My story . . .*, which together illustrate many of the features and opportunities that engineering offers and which the author has tried to highlight in the various chapters. You may find that reading the stories from different sections will help you to build up a rounded understanding of what a life in engineering offers. The case studies are drawn from past students, professional colleagues and friends and have been selected to be, in the main, examples of graduates in their early engineering careers (up to six years); Dick Elsy (Chapter 1), Geoff Mayes (Chapter 5) and Steve Dobson (Chapter 12) are examples of engineers well established in their careers (19 years plus). The case studies also include three entrepreneurs: Alastair Woodbridge (Chapter 4), Divya Surana (Chapter 15) and Gaston Chee (Chapter 17).

Chapter 15 deals with the process of finding your first graduate position. The sections on interviews and women engineers have been written by my colleague Dr Sophie Reissner-Roubicek based upon her research in this field. Her use of observations of real engineering graduate interviews of multinational students generated a number of useful insights which she shares. These are summarised with the aim of assisting all graduates, especially women, and interviewers as they seek to recruit more women into the profession.

Finally it must be acknowledged that a chapter of a few thousand words will never do full justice to the discipline that it purports to describe! Each chapter in Part 2 aims to provide a succinct summary from which you can focus in on those disciplines that interest you most. In the final chapter of the book you will find links to websites that can provide further detail. Many of the professional institutions, for example, publish materials for those interested in their discipline and some publish online versions of magazines or journals that provide current news and projects.

Profile: London South Bank University

London
South Bank
University

London South Bank University (LSBU) is one of London's largest and oldest universities.

Since 1892 LSBU has been providing students with industrially relevant, accredited and professionally recognised education. Over 25,000 students, from over 130 countries, choose to study with us each year.

Graduate prospects

We couldn't be prouder of the fact that we've been named the number one modern university in London for graduate prospects (Complete University Guide 2013).

Location

The university's location in the heart of SE1 is an asset and defines us. With excellent transport links we couldn't be better located and connected to City employers.

Why study engineering at LSBU?

You probably already know that engineering is one of the most well paid and secure of all professions. It's also one of the most exciting! We should know – we've been educating professional engineers for over 100 years.

Hands-on engineering

The amount of project-based learning that you'll do on an engineering degree varies from university to university. At LSBU we offer 'design-make-test' projects throughout the degree course rather than concentrating them all into your final year. This means that you'll adapt theoretical principles to solve real-world engineering problems very early on in your university career. This experience of delivering innovation makes you attractive to employers. Innovation is at the very heart of what an engineer does on a day-to-day basis. Engineers look for practical ways of making things better, more efficient, cheaper, safer, stronger, more resilient, quicker, more integrated and more effective. Our engineering courses will teach you first-hand how to develop these crucial skills and traits.

Prepared for modern engineering practice

In reality most engineers will find themselves working side-by-side in multidisciplinary project teams. One of the greatest professional assets that you can have is the ability to function well in this team set-up. That's why some of our modules are shared across all our engineering courses. These modules are about understanding the commercial priorities that shape engineering practice and problem-solving. Guest lecturers from world-renowned companies, such as Rolls-Royce, have lectured on these modules.

Engineering facilities

Our engineering workshops are the highest quality in the capital – thanks to the wide range of specialist equipment and space available for practical teaching. Highlights include the latest rapid prototyping facilities (including 3D printing), anechoic chamber and reverberation chambers, materials testing labs, mechatronics labs, electrical labs, building information modelling suite, industry-standard design and modelling software, and the Centre for Renewable and Efficient Building (CEREB) – which is London's first teaching, research and demonstration facility for low carbon technologies.

Case study

Illuminating engineering

The following summaries will help to give you an idea about what each engineering professional does. For more insight and detailed course information please explore our website lsbu.ac.uk, speak to our knowledgeable and friendly Course Advisors on **0800 923 88 88** or register to attend an on-campus Open Day at **www.lsbu.ac.uk/whats-on**.

BUILDING SERVICES ENGINEERING covers pretty much everything you can think of in a building – all the things that make it safe, comfortable and functional. This includes everything from making sure the floodlights at Old Trafford don't cut out halfway through the match to developing homes that can generate their own sustainable sources of gas and electricity.

Accreditation: Chartered Institute of Building Services Engineers (CIBSE)

CHEMICAL AND PETROLEUM ENGINEERING involves designing, developing, constructing and operating industrial processes to produce a huge range of products including oil and gas, pharmaceuticals, energy, water treatment, food and drink, plastics, and toiletries. It's an industry that is firmly focused on meeting the challenges of tomorrow; using the earth's resources as efficiently as possible to provide for the needs of future generations.

Accreditations: Energy Institute and Institution of Chemical Engineers (IChemE)

CIVIL ENGINEERING is all about designing, building and maintaining things that can change the world. Roads, bridges, canals, dams and buildings have all played an integral role in how the human race has developed. Civil engineers deliver a vital service in ensuring the safe, timely, well-resourced construction of a huge range of projects.

Accreditation: Engineering Council

ELECTRICAL AND ELECTRONICS ENGINEERING is all about designing, developing and maintaining electrical control systems. The impact of these engineers can be felt across many sectors. For example they provide lighting, heating and ventilation for buildings, they make sure that our transport networks run efficiently and safely, they help to power

Introduction

the manufacturing and construction industries, they develop wireless technologies and networks and they play a crucial role in the production and distribution of power.

Accreditation: Institution of Engineering Technology (IET)

MECHANICAL ENGINEERING can be seen and felt everywhere in the modern world. Mechanical engineers are employed across pretty much every sector you can think of – energy, transport, aviation, motor companies, robotics, pharmaceuticals, the marine industry – wherever you want to go, a degree in mechanical engineering can help you to get there.

Accreditation: Institution of Mechanical Engineers (IMechE)

MECHATRONICS ENGINEERING is an exciting branch of engineering that mixes mechanical, electrical, computing and control engineering. A classic example of the kind of product that is the result of mechatronics engineering would be an industrial robot such as those used on car assembly lines.

PRODUCT DESIGN ENGINEERING is all about innovation. Whether these new ideas are for a bicycle or a jet engine, a fountain pen or a super-fast laptop computer; product designers are the innovators who visualise tomorrow's products today, and then find ways to manufacture them intelligently and efficiently.

Accreditation: Institution of Engineering Designers (IED)

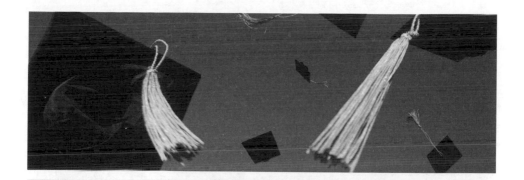

Profile: IGEM

IGEM (Institution of Gas Engineers and Managers) is a chartered professional membership body, licensed by the Engineering Council, for gas industry professionals operating in the UK and overseas.

Why join IGEM?

IGEM is a place where anyone working or interested in the gas industry can form positive connections to enhance their career and interest. In recent times, membership of a professional body has become a valuable asset and there has never been a greater need for keeping your knowledge and skills up to date to enhance your employability and career.

What can IGEM offer?

Members of IGEM enjoy deserved professional recognition throughout the UK and international gas industry, whether you are a manager, engineer or just starting out on the path to your future career. Our membership structure provides for entry at any level of qualification and at any stage in your career.

Students

Student membership is open to anyone pursuing a course of full-time or part-time study for a qualification that satisfies the educational base requirements likely to lead to corporate membership of the Institution.

Graduates

Graduate membership is open to anyone who holds an appropriate degree likely to lead to corporate membership of the Institution.

Progressing your career with IGEM

Membership of IGEM is fundamental to the recognition, representation and professional development demanded by successful engineers and managers throughout the gas industry.

IGEM can guide you towards your professional development and advise on how you can upgrade your professional membership and give you designatory letters after your name.

We can assign you with a mentor to help you achieve your personal goals. You can also become a mentor and we can support you on training workshops to enhance your mentoring skills. We can also advise you where to find the right course for your needs.

What's in it for me?

Sharing knowledge
You will gain access and be subscribed to our monthly journal with the latest industry and technology news. Members enjoy discounted rates to events, conferences and workshops hosted by IGEM throughout the year. Members have access and discounts to a range of IGEM standards which form the industry technical standards and codes of practice.

Networking
IGEM hosts a range of events and technical seminars which focus on specific industry technologies and allow engineering professionals to come together and offer expertise and advice on technical issues. Some of the events IGEM hosts every year include the Annual Conference and Engineering Update, the Gas Industry Awards and the Sir Denis Rooke Lecture. IGEM can give you the opportunity to meet fellow members at meetings and social functions.

Local Section involvement
IGEM has 10 District Sections, along with a Young Persons Network for those members under 35 years of age. The Sections are run by volunteers who organise a range of continuing professional development (CPD) events and networking opportunities.

Once accepted as a member, you will be allocated to a section based on your geographical location. This allows you to attend local events and meet other IGEM members plus discuss any topical issues and concerns. It also involves social events and visits to places of interest, such as power stations.

Gas International (Gi)
Gi journal is issued 10 times a year and provides you with top quality technical information from industry experts, designers and manufacturers. Items include new developments, changes in legislation, health and safety updates, installation guides, international and educational features and general industry news.

Letters after your name
Some of our grades of membership entitle members to use professional letters after their names e.g. John Smith CEng MIGEM.

As a member of IGEM, you will benefit from:

- international professional recognition
- accredited status as a competent and professional gas engineer or manager
- enhanced career prospects and business opportunities
- the opportunity to contribute to the rapidly evolving gas industry by active participation in the work of the Institution
- support for education, training and continuing professional development
- opportunities to meet fellow members of the profession at meetings and social functions, and through networking

- a growing network of technical and social contacts in gas engineering in the UK and overseas
- free subscription to our journal Gi
- discounted IGEM conferences and events
- discounted IGEM technical publications
- discounted conference and meeting facilities
- information and library services
- benevolent fund

Case study

Daniel Finley, Senior Engineer with MACAW Engineering Ltd, joined IGEM as a student member in 2006 while working for Advantica (DNV GL). Through his IGEM membership, Daniel is now a Chartered Engineer and chair of the Institution's Young Persons Network.

He said: 'When I started working I was still finishing my degree in Mechanical Engineering. A lot of the guys in the office were members of IGEM and I needed to join an institution to help with my professional development.

'When I graduated, I became an associate member and applied to become an incorporated member with the ultimate goal of becoming a Chartered Engineer. The application process was very slick and there was great communication throughout. I was able to deal with a real person, rather than being given a generic email address. You could always ring up and ask a question, even though I rarely needed to as the information I was supplied was always very clear. I had my interview on a Thursday afternoon and found out I had been successful the following Monday morning.

'One of the main things I enjoy about IGEM is receiving Gi every month and reading about the advances in engineering as part of my ongoing development.

'I've also got to know a lot of young engineers through the Young Persons Network and I would recommend IGEM to anyone who is currently working in the gas industry or just starting out in an engineering career.'

Professor Ghasem G Nasr, Head of the Engineering Research Centre and Director of Gas Engineering and Petroleum Engineering at the University of Salford, said: 'As a professor and director at the University of Salford, I believe that student membership with IGEM gives you the opportunity to keep up to date with industry movements and gain access to professional knowledge that will support the development of your future career path.

'Being a student member of IGEM is a great way to demonstrate you are committed to the industry as our future engineers and is a step in the right direction towards international professional registration.'

Part 1
Heading into engineering

1

Why work in engineering?

'Our profession, engineering, underpins the progress of humankind. For thousands of years, engineers have unlocked the natural resources of the earth for the benefit of humanity. Engineers have given practical application to scientific endeavour – driving economic growth and bringing billions out of poverty. In the future, it will be engineers using the earth's resources in new ways who continue that progress, solving the great challenges we face today.'[1]

'Overall, it is clear there is strong demand for engineering graduates in the UK with many employers expressing frustration at their inability to meet recruitment needs.'[2]

In addressing the question posed by this chapter – Why work in engineering? – one can focus on both the reasons why an individual might wish to work as an engineer and also how engineering is fundamental for our health, enjoyment, economy, society and so much more. If we pause to reflect we can recognise that every moment of our day depends on engineering in one form or another: water delivered through our taps is collected, purified and transported through the efforts of engineers; our clothes are manufactured, packaged and conveyed to us using engineered solutions; and our homes, vehicles, medicines and gadgets are all products of engineering. In this chapter we will therefore consider some of the reasons that motivate people to work in engineering in terms of their personal satisfaction and the reasons why as a nation and world we need engineers.

The attractiveness of engineering to the individual

Engineering is a rewarding activity that has the ability to generate tremendous personal satisfaction. It offers a career range that is diverse and in which each day brings new challenges to be solved. It is strange therefore that engineering has a legacy image that continues to misrepresent the truth of the engineer's daily activity, its importance for our daily and future lives, and the excitement and satisfaction that it can engender. This problem arises, in part, from the range of meanings attributed to the words **engineer** or **engineering** in the English language, such that it applies to everybody from the designer of spacecraft to the person called in to repair the family washing machine. This lack of clarity clouds the common understanding of the profession of engineering. In the public mind **engineering** conjures up images of working environments characterised by oil, grease and noise rather than the hi-tech, exceptionally clean environment of the electronics, precision engineering or high-end car manufacturer. In discussions with students, applicants and other interested young people, I am repeatedly struck by the fact that their interest frequently originates from a family member or friend who is a professional engineer and has provided them with a role model that belies the stereotype. Having a role model allows the individual to observe the challenges that engineering poses and the satisfaction that is derived from solving them and impacting positively on people's lives. This book describes the roles of chartered engineers or equivalent, focusing on their creativity in design and problem-solving.

Examples of the ways in which individuals have been attracted into engineering include a:

- fascination with some aspect of engineering: for example the agility and control of a fighter jet

- wish to improve the life of the world's poor: through, for example, the provision of clean water or better housing

- concern for the environment: perhaps through an interest in renewable energy or sustainability

- desire to make their 'mark' on the 'landscape': by constructing an iconic structure or manufacturing an iconic product

- motivation to create their own business or develop their design ideas

- recognition that their personal skillset is a good match for a career in engineering: for example they enjoy solving problems, enjoy mathematics and the science subjects relevant to their chosen engineering discipline.

There are other reasons for choosing to work in engineering, including the:

- diversity of roles between and within the different sectors of engineering

- joy of making a positive difference to others

- personal excitement and sense of fulfilment that creativity, innovation and problem-solving generate

- sense of having produced something that others value and use

- opportunity to engage in solving the big challenges that threaten the health and well-being of others including future generations

- satisfaction of working with other talented professionals in interdisciplinary teams

- opportunity for international travel

- future employment prospects (see *Employment Prospects* below)

- financial rewards that come with a successful engineering career (see *Salary* below).

Profile: University of Southampton

UNIVERSITY OF
Southampton

As an engineering student at Southampton you could be a part of our next world changing discovery!

Engineering at Southampton

The University of Southampton is one of the leading universities in the UK for engineering. Our academics are working at the forefront of their fields creating cutting-edge solutions to some of the world's biggest problems. The key technologies that drive today's global communication via the internet were developed by pioneers at Southampton.

Drawing on our world-leading research and taking advice from our industry partners, we tailor our degrees to ensure that our students are ready to tackle tomorrow's tough challenges today. We deliver a solid grounding of the fundamental principles of engineering and provide plenty of opportunities for students to put their theory to the test and gain essential practical experience.

There are over 200 specialist teaching and research laboratories and testing facilities dedicated to engineering across the university, including high-performance computers, 3D printers, wind tunnels, a high-voltage lab, flight simulators and towing tanks.

We offer a dynamic mix of lectures and seminars, practical lab sessions in our world-class facilities, site visits and project work.

We believe in helping our students gain the necessary experience for a future career, along with the skills to identify opportunities and make the most of them. At Southampton you will have the opportunity to broaden your options by meeting employers, getting involved in volunteering activities, work placements and much more.

Engineering degrees at Southampton:

Acoustical Engineering

Study sound and vibration; many advances today, from medical diagnosis to understanding climate change, are attributed to this discipline, as well as breakthroughs relating to audio and noise control.

Aeronautics and Astronautics

Aeronautics and Astronautics encompasses a broad range of disciplines in the field of aerospace engineering, with applications to the specification, design and construction of airframes, engines, satellites and other spacecraft.

Civil and Environmental Engineering

Civil and Environmental Engineering involves the planning, design, development and construction of solutions that enable the built and natural environment to be used in a sustainable way.

Computer Science and Software Engineering

Computer Science and Software Engineering gives you the skills to design software for demanding applications, while exploring new areas at the forefront of software development.

Electronic Engineering

Electronic Engineering explores electronics that can transform our world, including consumer electronics, computer processors and communication technology.

Electrical and Electromechanical Engineering

Electrical and Electromechanical Engineering covers elements ranging from power systems and electronics to computing.

Electrical and Electronic Engineering

Electrical and Electronic Engineering combines elements of both pure electronics and electrical engineering, and gives flexibility to study courses ranging from power systems to nano-scale devices.

Mechanical Engineering

Mechanical engineering is everywhere you look. It covers the design of machines, conversion of energy, manufacturing processes, medical engineering and microsystems technology.

Ship Science

Ship Science is the study of vehicles and structures that use the ocean for transport, recreation and harnessing marine resources.

Find out more: www.southampton.ac.uk/engineering/ug

Case study

Claire Gott, MEng 2010

I am a graduate structural engineer at WSP engineering consultancy based in Birmingham working on developments in the retail, health, education and commercial sectors across the UK.

I first became interested in how things work when I was 16. We went on a school trip to Tanzania to climb Mount Kilimanjaro and also built an orphanage. I liked how you could make a difference to people's lives in such a practical way.

My A levels were in Maths, Physics and Design Technology as well as an AS level in Geography. Initially I wanted to become an architect and I arranged two summer placements in England and Denmark but I realised how much I enjoyed using my maths, so I looked at opportunities in engineering instead.

The University of Southampton was one of only two places in the country that offered a degree in both Civil Engineering and Architecture. I went there on an open day and found it was one of the friendliest places I'd visited. My four-year Master of Civil Engineering with Architecture was an amazing course with excellent opportunities. I enjoyed the logic of engineering and the more creative subject of architecture and certainly developed the skillset I needed for my career.

One of the highlights of my time at university was co-founding Cameroon Catalyst in 2009 with Emily Hill. It is a student project which works with the Mosame Trust to support a village called Bambouti. I'm still working with the current Civil Engineering students to improve the lives of people in North East Cameroon. The latest project is to bring sustainable electricity to the villagers. On my most recent trip to Bambouti in August 2012, it was great to see first-hand what's been achieved to date.

During my studies, I had summer placements with Costain, working on three different projects: a critical care unit in a hospital, a hall of residence and a sewage treatment works. It was good to put into practice what I had learned at university. However, when I graduated in 2010, I decided to take a job with WSP. As well as working in structural design on multidisciplinary projects, I also get involved in my company's schools engagement plan 'Launchpad' and enjoy talking about careers in engineering to pupils. I've also gone back to my old school, The Kings School Worcester, to give presentations and stage a workshop. In fact one of the pupils, John Adeney, is now studying at the University of Southampton.

I am really enjoying my career and would recommend it to anyone who wants to use their skills and knowledge in practical hands-on work that is different every day.

Employment prospects

In the UK, some 5.4 million people work in engineering enterprises and it is estimated that 1.86 million new workers will be needed with engineering skills between 2010 and 2020.[3] As a STEM (science, technology, engineering and mathematics) subject, an engineering degree is highly prized by employers with 41% favouring them according to a recent CBI survey.[4]

The second quote at the top of this chapter is taken from a survey for the National HE STEM Programme in which it was noted that many employers' attempts to recruit sufficient engineering talent are frustrated. The short-term prospects are therefore very good and are further underpinned by the UK government's support for postgraduate courses in aerospace, etc. discussed in the appropriate chapters of this book.

According to EngineeringUK's analysis of the Higher Education Statistics Agency data, engineering also offers good opportunities for longer-term employment with 83.3% of engineers who graduated in 2007 in full-time employment in 2010, compared to the 72.3% average for all subjects.

Salary

Analysis of the 2007 cohort shows that in 2010 graduate engineers had the third highest median salaries at £28,000, below medicine/dentistry at £40,000 and veterinary science at £30,000. Below is a comparison of the annual mean gross pay for engineering-related professions based on a report by the Office for National Statistics. Median starting salaries for graduate engineers are typically within the range £24,000–£27,000.

- Managers in mining and energy: £68,860

- Research and development managers: £55,240

- Information and communication technology managers: £52,782

- Chemical engineers: £49,423

- Managers in construction: £48,634

- Electronics engineers: £45,558

- Electrical engineers: £44,558

- Mechanical engineers: £42,515

- Software professionals: £38,602

- Chartered surveyors: £38,139

- Design and development engineers: £37,982

- Civil engineers: £37,129

- Production and process engineers: £36,670

- Planning and quality-control engineers: £34,284

- Other engineering professionals: £39,397.

Source: 'The state of engineering' Office for National Statistics report, in EngineeringUK (2013)

The importance of engineering to the nation

'There are only three ways of creating wealth – you can dig it up, you grow it or you convert something to add value. Anything else is merely moving it about.'[5]

In the quotes at the top of this section and this chapter, both Sir John Rose and Lord Browne of Madingley highlight the importance of engineering and manufacturing for economic growth. Following the crisis in the financial sector that has enveloped much of the world, the UK government has recognised this truth and has introduced a raft of policies designed to bolster and encourage growth in the industrial landscape of the UK. Sir John Rose makes the point that an economy based upon the financial sector and consumerism alone is not generating wealth overall, but is simply moving it from one 'pocket' to another. Whilst some 'pockets' may accumulate more wealth, it will be at the cost of others. New wealth is generated by farming (growing it), mining a resource from the earth (digging it up) or using it to construct or manufacture (convert it) something that has additional value. Farmers depend upon machinery devised and manufactured by agricultural engineers and others and they may use fertilisers, insecticides and feedstock that are dependent upon chemical engineers. Mining engineers will design and operate mines and quarries to extract the raw materials, supported by engineers who provide machinery, ventilation, power, processing plant, etc. Most other engineering disciplines discussed in this book work to progressively convert the raw materials into valuable products and/or to protect the environment from our human activity.

However, for most people making a career choice based upon the general good for the nation's economy is not that attractive! Nevertheless, the fact remains that the quality of both our individual and societal lifestyles benefit from engineering as a strong wealth generator.

The global challenges demanding engineering solutions

As creative problem-solvers many engineers are inspired to apply their skills to the challenging problems that face our world. They recognise that engineers have a key role to play in working with other disciplines (sciences, economics, politics, sociology, health, etc.) to deliver solutions to those challenges.

The global population is expected to exceed 8 billion by 2050 placing an increasing demand on water, food, shelter and power, all of which require engineered solutions.[6] This growth and the associated needs are driving up the population of cities with the result that by 2050 the numbers of people living in cities vulnerable to earthquakes and cyclones will treble. This in turn, poses engineering challenges for the design of structures, services and coastal defences.[7] The impact that even the simplest engineering work can make is illustrated by the provision of a paved road or improved communication. Poor roads in Africa lead to almost as many fatalities as AIDS and have a similar impact on the local communities.[6] It is estimated that 1 billion people lack access to paved roads, with adverse consequences for their safety and economic potential,[6] whilst the introduction of high-speed mobile internet offers tremendous opportunities for improved communication, business and healthcare management.[8] Currently 783 million people cannot access safe drinking water[8] and 2.5 billion people have no access to proper sanitation facilities.[6] With the anticipated growth in population and the impact of climate change there are concerns that without improved engineering solutions for the sustainable management of our natural resources, the numbers of people deprived of clean water and sanitation will grow. Furthermore, we will struggle to meet the demand for water for human consumption, food and energy[9] resulting in increased migration, with some estimating that one person in every 45 of the world's population will flee the effects of climate change.[10]

The provision of infrastructure, communications, healthcare, security, clean water and energy for a global population that is both increasing in numbers and density can only be achieved through engineered solutions. Addressing these challenges will rarely be easy, but the business of engineering is to solve problems that do not always have a pre-determined, correct answer available in a book or on the internet.

My story . . .

Dick Elsy, Chief Executive of the High Value Manufacturing Catapult

I started my career as a sponsored student with Climax fork lift trucks and when I graduated in 1981 I joined the company to work as a development engineer. In 1983 I moved to Land Rover to work as a vehicle development engineer. Promotion followed quickly, achieving my first management position leading a minor vehicle programme in 1985. The big break came when I was selected as part of a breakthrough team to develop the original Discovery vehicle. We delivered the Discovery vehicle programme and I was promoted to chief engineer for Discovery in 1989 and at 29 years of age was the youngest executive in the Group at that time. Two years later, I joined the Land Rover Board with responsibility for all Discovery and, later, all Defender engineering. During this time I developed a new product strategy for Land Rover that included the vision for a new vehicle – Freelander. I was given the rare opportunity to develop this new vehicle through to production. After delivering Freelander, an award winning vehicle that generated £10 billion for the UK economy, I led a major vehicle programme for BMW in Munich.

In 1999 I joined Jaguar as Engineering Director, responsible for leading a 2,500 strong engineering team to deliver all product actions for Jaguar Cars, introducing innovations such as all aluminium body construction.

I moved to Torotrak Plc. as Chief Executive in 2003 and led a turnaround of this business which develops novel traction drive transmissions (a form of gearless gearbox). Shareholder value grew by some 3.5 times in my tenure, and Torotrak was confirmed as the second best performing 'small-cap' company on the FTSE in 2011.

My current role is Chief Executive of the High Value Manufacturing Catapult, a government programme to accelerate innovation to commercial reality and grow the UK's manufacturing capability. My job is to develop the strategy to support small and large manufacturing business for sustainable UK growth. We do this through working with these businesses to help to develop innovative manufacturing processes which have the ability to add billions of pounds to the UK economy.

What excites me about engineering? I love the principle of the power of wealth creation through engineering. I have personally been involved in major programmes which have earned the UK sums in the range millions to billions of pounds. Engineering adds value in a tangible way that many other careers cannot match. I also love the creativity and the fact that you can use your skills to solve not only engineering problems, but business ones as well.

In my 30-year career, I don't think that there has been a better time to be an engineer. The world has some big challenges facing it: need for growth, energy security and supply, climate change, etc. The solutions to these challenges will always lie in technology (the alternatives are to regress) so the engineer is central to the economic and environmental well-being of the planet – a role which is not surpassed in influence. I have had a fantastic time being an engineer and have been given extraordinary responsibilities and challenges. My advice to anybody considering a career in engineering is: 'Grasp every opportunity given to you – you will be rewarded for your efforts!'

2

Engineering professional, statutory and regulatory bodies (PSRBs)

'Global competence is among the new skills and abilities needed for all graduates to live and work knowledgeably and comfortably in a transnational economy and global society, especially for engineers.'[11]

For those considering working in engineering this chapter offers an overview of the various organisations that will influence your education and training and provide you with the means to professional recognition within your chosen career path. These organisations also facilitate the means for qualified professionals to work internationally and have their qualifications recognised in other countries. A number of these bodies publish reports that may provide useful insights into the nature of the engineer's work and the potential for them to contribute to solving the global challenges that loom large.

At first glance the UK engineering profession can seem very confusing in terms of the number of agencies that regulate the various aspects of professional practice. However, on closer inspection it becomes apparent that each agency performs a unique role in support of the whole. They provide a structure that covers everything from promotion of engineering as a career, through specification of the necessary education and training required to demonstrate competency as an engineer, to professional registration and assurance of standards. They also act as agents to influence public policy and understanding.

Engineering is a global industry and whilst the scope of this book is primarily focused on the UK we will make reference to some of the international regional organisations and the key international frameworks that facilitate the mobility of engineers between countries.

Within the UK, professional registration is the responsibility of the Engineering Council UK supported by the various professional institutions. In this chapter we will not discuss each of these institutions in detail for space does not allow this. However, we will make reference to key professional institutions in those chapters dealing with the different branches of engineering, and a full list of professional engineering institutions is provided on the ECUK website (www.engc.org.uk/about-us/our-partners/professional-engineering-institutions).

Engineering Council UK (ECUK)

The number of professional engineering institutions in the UK demands an organisation that will ensure consistency of standards across the profession and is capable of representing the interests of engineers to government, society and the media. The ECUK is the current holder of that role. As the regulatory body for the engineering profession it is responsible for maintaining a register of engineers in four grades, each determined by the nature of the education and work that the individual has completed. The four grades are listed below.

- **Chartered engineer (CEng):** a chartered engineer is characterised by their ability to solve new engineering problems through the application of advanced analytical, design and/or management skills. Their approach may demand the development or application of new technologies or techniques to achieve an efficient and cost-effective solution. They may pioneer new engineering services and management methods. Chartered engineers will require strong mathematical and modelling skills matched by creativity in design and acute business acumen. Their role requires excellent communication and interpersonal skills to support their dealings with customers, suppliers and their work colleagues. They will frequently work in teams and be required to demonstrate technical and commercial leadership.

The title CEng is protected by civil law and is recognised internationally as demonstrating engineering competence. Chartered engineers are variously engaged in technical and commercial leadership and possess effective interpersonal skills.

- **Incorporated engineer (IEng):** like chartered engineers the incorporated engineer will need strong interpersonal skills to support their work in teams and with other stakeholders. Incorporated engineers will manage new and existing technologies and processes. They will also contribute to the design, development, deployment and decommissioning of products. Incorporated engineers may be engaged in technical and commercial management roles.

The title IEng is awarded in recognition of the individual's demonstrated competence, commitment, skills and experience.

- **Engineering technician (EngTech):** engineering technicians may have supervisory or technical responsibility and will apply proven techniques and processes to the solution of practical engineering problems. Like all engineers they are expected to adopt safe systems of working. They will have skills within defined fields of engineering and will use these to contribute to design, development, manufacture, commissioning, decommissioning, operation or maintenance of products, equipment, processes or services.

- **ICT technician (ICTTech):** the grade of ICT technician recognises the role of those working with information communications technology equipment and applications, including the installation, operation and maintenance of hardware, software or systems.

The ECUK maintains the register of engineers by stipulating the expectations for professional competence, ethical conduct and social responsibility expected of all registered members. It publishes the standards expected in terms of professional competence in the form of a specification document, The UK Standard for Professional Engineering Competence (UK-SPEC). This document lists the generic competences expected of each engineering grade and is used to inform the accreditation process for engineering degrees and is thus the measure against which they are judged. The ECUK licenses 36 professional engineering institutions to accredit courses and assess individuals on its behalf.

UK professional engineering institutions

Arguably the primary unit of the Professional, Statutory and Regulatory Body (PSRB) is the individual professional engineering institution (PEI) established to support a particular branch of engineering. These are a family of organisations committed to promoting good practice through the sharing of information and education within their disciplines. Frequently a single PEI will include a number of specialist sub-groups.

The PEI family tree is complicated and the present structure has resulted from alliances and mergers formed over the last three centuries. The original PEI was the Corps of Engineers formed in 1717 to reflect the then interest in military engineering. As new technologies have developed they have spawned new institutions to support their use and development. So in 1818 the Institution of Civil Engineers was formed to represent the growing engineering activity without a military interest, reflecting the growing activity in construction of ports, waterways and public health systems. Later, in 1847, the railway era inspired the formation of the Institution of Mechanical Engineers and in 1871 the Institution of Electrical Engineers was launched, later to become the Institution of Engineering and Technology. An excellent example of the dynamic nature of institutional merger is that of the Institute of Materials, Minerals and Mining (IOM3) which can trace its roots to 26 groups dating as far back as the mid-nineteenth century and an assortment of regional mining institutes.

The PEIs act individually or in partnership to recognise the professional competence of individuals tested against the ECUK's specification. Each institution interprets and applies the specification to their discipline. The process of establishing professional competence is streamlined through the accreditation of academic courses and industrial training schemes within companies.

In pursuit of their educational objectives most PEIs will organise professional meetings and presentations and publish specialist journals and books in their discipline. Many will also be active in trying to attract young people and others into their industries. Those interested in moving into a career in engineering will find that PEI webpages will often contain useful information about training and careers.

EngineeringUK

In recent times engineering as a profession has needed to actively promote itself as an interesting and rewarding career route. EngineeringUK is the lead organisation for raising public awareness of engineering and promoting it as a career. This is achieved through outreach to schools and the Big Bang Fairs, which are the largest celebration of science and technology in the UK, supported by the spin-off The Big Bang Near Me regional events.

EngineeringUK also compiles a useful annual snapshot of engineering activity and examines the state of the academic supply chain that feeds degree courses and graduate recruitment, which it publishes in its annual statistical report. This report is primarily targeted at the engineering community, but those considering a career in engineering can find up-to-date information on salary structures, demand for graduates and skills shortages.

Royal Academy of Engineering

The Royal Academy of Engineering (RAEng) is a registered charity with the aim of advancing and promoting excellence in engineering. Its role differs from that of a PEI because it takes a holistic view of engineering rather than focusing on one particular discipline. Its membership is also restricted to the most accomplished of engineers. The RAEng publishes the magazine *Ingenia*, which provides an excellent insight into the current trends and developments within engineering. *Ingenia* can be viewed online at www.ingenia.org.uk. The RAEng commissions research in association with other funders and also organises a range of conferences and courses. A number of reports have focused on the needs of the profession in terms of education, for example the RAEng has published reports on engineering systems, ethics and the expectations of employers of graduates.

The RAEng has sought to promote engineering as a career to young people through the establishment of its taster courses, such as Headstart and Dragonfly, which are run in conjunction with the Engineering Development Trust.[12]

The Engineering the Future Alliance

This organisation is linked to the Royal Academy of Engineering and was established with the aim of supporting policymakers, such as government departments, parliamentary committees and agencies, in their understanding of engineering and the role it plays in the economy and society in general. The Alliance undertakes policy studies and other activities to support national policy-making that has an engineering or technology dimension.

Education for Engineering (E4E)

Education for Engineering is hosted by the Royal Academy of Engineering and takes its membership from the professional engineering institutions, Engineering Council and EngineeringUK. E4E represents the breadth of the professional engineering sector when advising the government and devolved Assemblies on matters of education and engineering skills policy.

Engineering Development Trust (EDT)

Engineering is always seeking ways to attract talented young people to work in the industry. To achieve this there are schemes designed to excite interest and to provide a taste of what engineering can offer. For young people interested in exploring engineering as a potential career, the EDT provides an extensive range of STEM enrichment activities for ages 11 years and upwards. By working in partnership with education and industry, EDT offer projects, courses and placements that provide experience and develop the relevant skills. Details can be found through their website,[13] but schemes of particular note include residential taster courses hosted by universities (Headstart) and events designed to support under-represented groups. EDT also organises the Year In Industry scheme, providing paid work placements for students taking a gap year before or within a university course. The fact that employers value graduates who have undertaken such placements is discussed in Chapters 14 and 15.

International Engineering Alliance

The International Engineering Alliance promotes mutual recognition of professional engineers and thus mobility across its signatory regions and countries. The Alliance is 'home' to a number of agreements.

- **The Washington Accord:** this international agreement provides for the mutual recognition of courses accredited by the signatory accrediting organisations, such as the ECUK. This provides for the international mobility of graduates of courses accredited as meeting the academic requirements for chartered engineer status.

- **The Sydney Accord:** this accord is the equivalent to the Washington Accord and provides for recognition and mobility of engineering technologists and incorporated engineers.

- **The Dublin Accord:** this provides for the international recognition of engineering technician qualifications. It currently has four signatories (Canada, Republic of Ireland, South Africa and the UK) and both New Zealand and the United States have provisional membership with the intention of achieving signatory status.

A further three agreements focus on the recognition of professional competence based on the individual rather than the nature of their qualifications. The purpose of these agreements is that an individual who is recognised as meeting the professional competencies in one country should be facilitated in registering as an engineer in a second country with a minimum of further assessment, focused on local knowledge rather than broader professional competence. These agreements are:

- The APEC Engineer Agreement

- The Engineers Mobility Forum

- The Engineering Technologist Mobility Forum Agreement.

European Network for Accreditation of Engineering Education (ENAEE)

The ENAEE seeks to 'build confidence in systems of accreditation of engineering degree programmes within Europe and to promote the implementation of accreditation practice for engineering education systems in Europe.'[14] In particular, the ENAEE administers EUR-ACE as a European recognition of engineering degree programmes at both bachelor and master's level.

World Federation of Engineering Organisations (WFEO)

Founded under the auspices of the United Nations Educational, Scientific and Cultural Organisation (UNESCO), the World Federation of Engineering Organisations is a non-governmental international group that represents some 15 million engineers from over 90 nations. WFEO declares its mission to encompass:

- representing the engineering profession internationally

- enhancing engineering by promoting best practice internationally

- seeking to advise on policy choices in areas such as:

  ○ fostering socio-economic security

  ○ sustainable development

  ○ poverty alleviation throughout the world.

European Federation of National Engineering Associations (FEANI)

The European Commission recognises the FEANI as the voice of the engineering profession from 32 countries in Europe and representing in excess of 350 engineering groups with national status in their home countries. The FEANI also enjoys consultative status with UNESCO, UNIDO and the Council of Europe. The

Federation exists to promote and protect the status, role and place of professional engineers within Europe and to promote their ability to work worldwide.

Commonwealth Engineers' Council (CEC)

With the mission statement 'CEC advances the science, art and practice of engineering for the benefit of mankind',[15] this organisation works to nurture the professional development of engineers within the Commonwealth through education, training and technology transfer. The CEC represents the views of the engineering profession to government departments and heads of government within the Commonwealth.

Federation of Engineering Institutions of Southeast Asia and the Pacific (FEISEAP)

This international not-for-profit professional organisation represents and supports the work of engineers in Southeast Asia and the Pacific region. As with most associations of this type, the focus is on education, professional development, mobility of engineers between countries and providing support to groups and government on engineering issues pertinent to the region.

ASEAN Federation of Engineering Organisations (AFEO)

AFEO represents the engineering institutions and organisations of ASEAN countries with the purpose of establishing a regional baseline standard for the engineering profession and thus promoting mobility of engineers within the member countries.

3

Routes into engineering

'Opportunities exist for professional engineers at all levels and engineering is one of the few career areas where there are clear professional progression routes through work-based learning.'[16]

As we have seen in Chapter 2 the Engineering Council UK (ECUK) maintains a register of engineers in four grades, each recognising that the individual has achieved certain academic and practical knowledge and skills:

- chartered engineer (CEng)

- incorporated engineer (IEng)

- engineering technician (EngTech)

- ICT technician (ICTTech).

The routes into engineering must therefore provide a means of demonstrating that the required competencies have been acquired. To understand these routes we must first relate them to the different levels of qualification. This is complicated by the fact that Scotland operates a different scale to those of England, Wales and Northern Ireland. A useful comparison is published by the Office of the Qualifications and Examinations Regulator, Ofqual.[17] Each of the grades above is related to the qualification frameworks in Table 1 in which the numbers in the first two columns give the level of the qualification. Chartered engineer status is shown as requiring an educational component equivalent to a master's degree, which is NQF level 7 or SCQF level 11.

TABLE 1: National Qualification Frameworks (NQF)

UK excl. Scotland	Scotland SCQF	Indicative qualifications	Professional level[†]
8	12	Doctoral degrees (PhD; EngD) Vocational qualifications level 8	
7	11	Integrated master's degrees (MEng) Master's degrees (MSc) Postgraduate Diploma (PGDip) Postgraduate Certificate (PGCert) NVQ level 5 SVQ level 5 Vocational qualifications level 7	Chartered engineer (CEng)
6	7–10	Bachelor's degrees (BEng; BSc) Graduate diplomas (GDip) Graduate certificate (GCert) Vocational qualifications level 6	Incorporated engineer (IEng)
5		Foundation degrees (FdEng) Diplomas of Higher Education (DipHE) Higher National Diplomas (HND) NVQ level 4 (England) SVQ level 4 Vocational qualifications level 5	Engineering technician (EngTech)
4		Advanced Highers (Scotland) Scottish Baccalaureate Higher National Certificates (HNC) Certificates of Higher Education (CertHE) NVQ level 4 (Wales) SVQ level 3 Vocational qualifications level 4	

UK excl. Scotland	Scotland SCQF	Indicative qualifications	Professional level[†]
3	6	GCE AS and A level Highers (Scotland) Welsh Baccalaureate Advanced Advanced Diploma (England) Progression Diploma (England) Apprenticeships Framework (Wales) NVQ level 3 SVQ level 3 Vocational qualifications level 3	ICT technician (ICTTech)

2	5	GCSEs at grade A*–C
		Welsh Baccalaureate Intermedi-
		    ate
		Higher Diplomas (England)
		Foundation Apprenticeship
		    Framework (Wales)
		National 5 (Scotland)
		Intermediate 2 (Scotland)
		Credit Standard Grade (Scotland)
		NVQ level 2
		SVQ level 2
		Vocational qualifications level 2	

1	4	GCSEs at grade D–G Foundation Diplomas (England) Welsh Baccalaureate Foundation National 4 (Scotland) Intermediate 1 (Scotland) General Standard Grade (Scotland) NVQ level 1 SVQ level 1 Vocational qualifications level 1	

[†] These levels are indicative of the academic component and their achievement also requires demonstration of practical competencies and experience appropriate to the grade. There are routes that may be followed to allow progression from lower points on the framework.

Each ECUK grade represents a 'destination' on the 'academic map' of Table 1 and the routes available to each destination will depend upon the starting point. This chapter is therefore structured to reflect this. For simplicity we will refer to the Framework levels for England, Wales and Northern Ireland. Levels on the Scottish Framework may be translated using Table 1.

The educational landscape in the UK, especially in terms of type of provider, is both dynamic and increasingly diverse at NQF levels 1–3 leading up to university admission, making any comprehensive summary here difficult. The traditional route followed GCSEs (NQF 2) and AS and A level qualifications (NQF 3), but today schools and colleges may also offer baccalaureates, diplomas and other alternatives or combinations of qualification. The reader should therefore view the discussion as indicative of the options available to them and follow the links in the final chapter, 'Further Resources and Information', for further details of particular qualifications. All universities will publish guidelines about their requirements for specific awards or will provide information upon request.

Profile: Institute of Acoustics

The Institute of Acoustics (IOA) is the UK's professional body for those working in acoustics, noise and vibration.

Institute of Acoustics' Diploma in Acoustics and Noise Control

Formed in 1974, the IOA has some 3,000 members from a rich diversity of backgrounds, with engineers, scientists, educators, lawyers, occupational hygienists, architects and environmental health officers among their number.

The Institute's Diploma in Acoustics and Noise Control, which has been offered since 1975, is the leading specialist qualification for the professional practitioner in acoustics. There are a number of benefits of obtaining the diploma.

- It provides participants with an in-depth understanding of the theory, principles and practical applications in relevant aspects of acoustics and noise control.

- It's accepted in partial fulfilment towards acoustically related master's degrees by partner universities.

- It provides a route to achieving corporate membership of the IOA and for CEng status.

Who should attend?

The course is aimed at all those who wish to be professionally engaged in the fields of acoustics and noise control, including engineers, architects, acoustic and environmental consultancies and local authority health practitioners, scientists and technicians.

It is usually studied on a part-time basis over one year and is available at accredited centres in the UK and Ireland. A tutored distance learning option involves the distribution, to a timetable, of learning-support material and exercises. This is supported by a programme of small group tutorials (which can also be joined by video-conferencing for those unable to attend) and two-day intensive residential laboratory work.

This format should be of particular interest to people in regions where centre-based courses are not available or who live abroad. It is also possible to study for individual modules of the diploma.

The following modules are offered:

- general principles of acoustics (mandatory)

- laboratory module (mandatory)

- environmental noise, prediction, measurement and control (specialist module – optional)

- regulation and assessment of noise (specialist module – optional)

- building acoustics (specialist module – optional)

- noise and vibration control engineering (specialist module – optional)

- project module (mandatory).

Except for the project and laboratory modules, assessment is through coursework and written examinations. The laboratory module is assessed through a laboratory notebook and three written laboratory reports.

The diploma is awarded for the successful completion of the general principles of acoustics module, the laboratory module and any two specialist modules and the project module.

What entry requirements do I need?

Entry requirement is a first degree in a relevant science-based subject, such as physics, engineering or environmental health. However, the IOA encourages an APEL-based open access policy as befits a vocationally focused course: a key factor being the ability and enthusiasm of students to complete their studies.

Further information on accredited centres, course tutors and the distance learning option may be obtained from the IOA website www.ioa.org.uk/education-and-training/ or by phone 01727 848195 or email: education@ioa.org.uk

Case study

The Institute of Acoustics is one of several bodies licensed by the Engineering Council to register members as professional engineers.

Getting chartered: the route to an internationally recognised award

Through its Engineering Division, suitably qualified and experienced engineers may gain this internationally recognised award. Here are case studies of three members who have recently been accorded this status:

James Hill, AAF, IEng

James graduated in 2008 from the University of Salford with a BSc in Acoustics. He then joined AAF as a trainee acoustic engineer and has since progressed to his current role as acoustic engineer. In addition he is the UK representative for the European Acoustics Association Young Members' Group.

'As I work primarily in an engineering environment and deal with clients from all over the world, when the opportunity arose through the institute to achieve IEng status it seemed an obvious thing to do,' he said. 'The process itself helped me to understand the requirements and gave me the motivation to get my professional records in line, which will hopefully help me in achieving full CEng status further down the line.'

Dr Ning Qi, Doosan Power Systems, CEng

Ning graduated from Tsinghua University in Beijing, China, with a BSc in Process Control Engineering and a MEng in Multi-Phase Flow Measurements. He obtained his PhD in Acoustics from the University of Liverpool in 2000.

Ning worked in different academic institutions before joining Doosan Power Systems in 2007. He is now a principal engineer working in a wide range of noise and vibration projects across the company business sectors, including power generating, petrochemical, and oil and gas industries, to develop appropriate solutions to engineering problems.

'I started my application for CEng status as a response to the company's "Get Chartered" campaign,' he said. 'But my drive for becoming registered as a CEng was to receive visible acknowledgment of my varied experience and for my own sense of achievement. I also believe that it helps in engaging with other professionals.'

Mark Scaife, WSP Middle East, CEng

'As I have a non-accredited degree (audio technology, University of Salford), I followed the individual report route to registration,' he said. 'This involved gathering a range of project examples together and annotating them to demonstrate where the various competencies are met.

'I found it useful to provide covering notes to the technical reports to provide information on some of the commercial aspects of my role, that are not necessarily picked up in a consultancy report. The input from Peter Wheeler (IOA Engineering Manager) prior to the interview was also extremely helpful.

'Becoming a chartered engineer is becoming increasingly important in the Middle East where I am based, with clients demanding senior members of the design team not only have relevant experience but are chartered as well.

'Aside from that, I am extremely proud that I have been thoroughly assessed by my peers and that I "made the grade". I thoroughly recommend it.'

Case study

Michael Wright studied BEng Mechanical Engineering at Oxford Brookes and completed an industrial placement with Oxford Optronix as part of his degree.

Michael started his time at Oxford Brookes with a Foundation year in Engineering, to prepare him for a BEng in Mechanical Engineering. He needed to learn transferable skills in sensors, instruments and control, for a proposed future career manufacturing racing kayaks and paddles, so in the third year of his BEng programme he obtained a one-year industrial placement with Oxford Optronix, a locally based company specialising in the development and manufacture of sophisticated instrumentation for clinical medicine and the life sciences. This placement year also helped Michael to develop his time management skills which came in useful in the final year of his degree!

Michael says that studying engineering is hard but very rewarding. The courses are 'full on', but you learn so many useful skills and the experience as a whole fits you for working in the diverse area of engineering. Michael says, 'The way you think changes – you become more analytical. Having a degree in engineering holds a lot of weight with employers.'

During his placement year Michael had hands-on experience of quality management systems, gaining valuable insight into how to deal with the bureaucracy surrounding this essential part of everyday practice in modern engineering. He also managed a new project and saw his designs through to prototype and eventual product launch in Chicago, picking up useful skills in electronics on the way.

The Foundation year in Engineering is a valuable route into engineering study at degree level. It is especially useful for applicants whose A level grades in Mathematics and Physics don't match what's needed for a degree or whose A levels are not in those subjects or who don't hold a comparable qualification. As Michael says, the course teaches you the specific mathematics and physics you need for the degree course, making it the ideal springboard for higher study. Indeed, he says it is the reason he decided to study at Brookes.

Michael really enjoyed his time at Oxford Brookes. The degree course has 'lots of contact hours and the teaching is fantastic', he says; he was particularly impressed with the staff's passion for their subject and the time they gave to supporting him.

'Working in a placement allows you to see what is necessary to your aspirations,' Michael commented. He graduated this summer and is returning to Oxford Optronix as a fully-fledged member of staff. We wish him every success.

Starting NQF level 1

Most students will undertake GCSEs or their equivalent with the aim of achieving the progression level required for NQF level 3 study. The mix of subjects studied should allow for advanced study of mathematics and the physical sciences. Future entry to a university degree course will normally require a range of subjects including English language. Design and technology subjects will give an insight into the design process and may offer relevant practical experience with tools.

An alternative to GCSEs exists in the form of the Diploma in Engineering which is an applied qualification developed with employers to provide subject specific skills and knowledge alongside employability skills and a self-managed project. Unlike GCSEs the diploma can be extended to complete NQF level 3. The awards available are:

- Foundation: equivalent to 5 GCSE grades at D–G (NQF level 1)

- Higher: equivalent to 7 GCSE grades at A*–C (NQF level 2)

- Progression: equivalent to 2.5 A levels (NQF level 3)

- Advanced: equivalent to 3.5 A levels (NQF level 3).

For those with the ambition to study for a university degree the Advanced Diploma in Engineering should be supplemented with the additional qualification: OCR Level 3 Certificate in Mathematics for Engineering.[18]

In Scotland, students study for the Nationals, which replace the previous Standard Grade examinations and Intermediate 1 and 2, to permit progression to Highers.

For those eligible for employment, options exist to undertake an apprenticeship and/or to complete vocational qualifications (e.g. NVQ and SVQ) that recognise competency in the workplace. Apprenticeships are discussed below.

NQF level 3

A range of qualifications exists that provide a route to university admission on an engineering course:

- AS and A level

- Scottish Highers and Advanced Highers

- Welsh Baccalaureate

- Scottish Baccalaureate (Science)

- International Baccalaureate.

The subjects studied at the most advanced level should normally include mathematics and physics, but some disciplines may have additional requirements, for example chemical engineering degrees will require chemistry.

For those who may have studied subjects that do not meet the entry requirements there are options such as foundation programmes, Year 0 entry to degree courses or additional support that allow capable candidates to redirect their studies to engineering.

Some may wish to enter an apprenticeship in which they combine academic study with work experience. Admission to these schemes may be managed through the company offering the apprenticeship rather than the university offering the academic study.

NQF levels 4 and 5

Post-NQF level 3 a number of options exist to progress your engineering career through entry to a:

- Bachelor (BSc or BEng) or integrated Master of Engineering (MEng) degree course

- Higher National Certificate (HNC) or Diploma (HND) course

- Foundation degree (FdEng) course.

The first two may be studied full or part time according to the institution offering the degree. Some large employers, such as Jaguar Land Rover, have designed degree programmes in partnership with universities that allow their employees to gain a degree whilst they work.[19]

Foundation degrees are typically part time with employer support for the practical training. HNC, HND and Foundation degrees allow for progression to a degree course with advanced entry related to the NQF level achieved — however, progression is at the discretion of each university and this may restrict the individual's choice. Progression will be dependent upon the university having space on the degree course and there being a good curriculum match with the previous studies. Some universities may also impose additional requirements in terms of modules studied within the prior qualification, especially in relation to engineering mathematics.

An appropriately accredited HND or Foundation degree will meet the academic component for ECUK registration as an engineering technician.

NQF level 6

Degrees at NQF level 6 may be accredited as:

- meeting the educational requirement for ECUK registration as an incorporated engineer, or

- partially meeting the educational requirement for ECUK registration as a chartered engineer and therefore requiring further learning, as discussed below under NQF level 7.

It should be noted that professional engineering institutions (PEI) restrict accreditation to honours degrees and in some cases do not recognise Third Class honours degrees.

For graduates from closely related degrees, such as mathematics, physics, chemistry and environmental science, it is possible to find niche employment within the engineering sector or to 'convert' to engineering by means of one or more of the following:

- graduate diploma: typically used to bridge from one degree to the desired branch of engineering

- MSc: providing either a conversion route or advanced engineering education

- PhD: providing research training in a field of engineering that uses the first degree expertise

- Undertaking a bespoke set of academic modules to satisfy a PEI's requirements.

Some graduates may determine to change profession by undertaking a new undergraduate degree in engineering, but this is a costly endeavour in the present fee structures.

NQF level 7

The educational requirement of the criteria for CEng status is fully met by gaining an accredited NQF level 7 qualification or an equivalent. The MEng degree provides this, but for those with a first degree accredited as partially satisfying the educational requirements and who need to undertake further learning, the options include:

- an accredited MSc

- employer managed further learning: a recognised and structured scheme in which the employer provides the missing educational content

- the Engineering Gateways programmes: providing a 'flexible work-based learning "escalator" that allows progression to

professional registration without the need to leave work'
(www.engineeringgateways.co.uk)

- self-managed further learning: used where the employer does not offer a recognised scheme

- technical report route: allowing demonstration of the academic knowledge gained by means other than formal academic study, including work experience. Each PEI provides guidance on how to submit an application via this route.

Apprenticeships

The Institution of Engineering and Technology (IET) opinion report, 'Furthering Your Career', concludes that, in the face of increased undergraduate fees, the apprenticeship route may grow in popularity: '. . . a balance of theoretical and on-the-job training is required for engineers, with advantages to both a graduate entry route and an apprenticeship route, the overwhelming opinion was that, particularly with the increasing cost of degrees, the practical skills offered by apprenticeships could be a more attractive option for future engineers.'[20]

The report also noted: 'However, it was felt by many that academic qualifications came into play more as you reach senior positions and require more theoretical-based business and leadership skills, whereas professional qualifications were more important for hands-on roles.'

Apprentices study whilst working alongside experienced staff to master practical job-related skills. Academic study is normally provided via day-release study for a recognised qualification. In the UK apprentices must be 16 years of age or older and not in full-time education. Apprenticeships take up to four years to complete depending on the level, industry sector and the individual's ability.

Apprenticeships are available at three levels:

- Intermediate (Foundation in Wales): leading to a NQF level 2 vocational qualification

- Advanced: a NQF level 3 vocational qualification

- Higher: vocational qualifications at NQF level 4 and potentially a knowledge-based qualification such as a Foundation degree.

Advice on apprenticeships and participating companies is available from the websites for the following organisations:

- National Apprenticeship Service (www.apprenticeships.org.uk)

- Apprenticeships in Scotland (www.apprenticeshipsinscotland.com)

- Welsh Government (wales.gov.uk/apprenticeships)

- notgoingtouni.co.uk (www.notgoingtouni.co.uk).

My story . . .

Sarah Chen, mature student at University of Warwick pursuing a change of career

I came to the UK to study for an MSc in Virology at the London School of Hygiene and Tropical Medicine and followed this with a PhD in Immunology at Oxford University. I later switched to teaching to more easily accommodate my young family. I taught biology, chemistry and physics in schools in Warwickshire and became Head of Science for four years. I found that I no longer enjoyed teaching and decided to take the plunge and change to my lifelong dream career. I returned to university to study civil engineering because I had always wanted to study engineering but my parents did not approve of the idea and refused to support me.

My experience of returning to undergraduate study has been 'the best of times and the worst of times'! On the positive side I:

- really enjoy the course

- am surrounded by nice course mates and interesting people with great social and cultural diversity

- find my department helpful and accommodating in meeting my needs

- have accessed wonderful experiences and opportunities through joining clubs and societies and volunteering, including working on hydro projects in Uganda.

As a mature student I have found some aspects of returning to education difficult:

- supporting myself financially

- balancing family and course commitments

- settling back into a learning environment

- finding I had forgotten all my maths!

- I find it takes longer to learn and remember new material.

I am attracted to engineering by the opportunities it offers to influence the environment and policies; in particular in the areas of sustainability, renewable energy and development.

Postscript

The demand for engineers at all grades is such that the industry, supported by government, has endeavoured to provide routes into and through engineering grades at all levels. For some the choice will seem overwhelming and potentially confusing. Useful diagrams illustrating the range of routes are available for England,

Northern Ireland, Scotland and Wales on the Tomorrow's Engineers Website (www.tomorrowsengineers.org.uk/Careers_resources/Individual_resources).

Ultimately, the individual must judge:

- where they are currently placed on the NQF or SCQF levels in regard to engineering – their starting point

- the professional grade or role that they aspire to achieve – their destination

- the routes available to them – dependent upon their location (national variations), existing qualifications and grades achieved

- their personal preference in terms of the balance between employment and study.

4

Engineering knowledge and skills

'Professional engineering is not just a job – it is a mind-set and sometimes a way of life. Engineers use their judgement and experience to solve problems when the limits of scientific knowledge or mathematics are evident. Their constant intent is to limit or eliminate risk. Their most successful creations recognise human fallibility. Complexity is a constant companion.'[21]

'It is this combination of understanding and skills that underpins the role that engineers now play in the business world, a role with three distinct, if interrelated, elements: that of the technical specialist imbued with expert knowledge; that of the integrator able to operate across boundaries in complex environments; and that of the change agent providing the creativity, innovation and leadership necessary to meet new challenges.'[22]

What makes an engineer? This question will generally invoke an answer that argues for one or more of the following elements:

- applied mathematics: building a model or simulation of the problem that permits an optimal solution to be arrived at

- design: highlighting the creative and innovative facet of engineering

- business: taking your design or expertise to market, either as your own company or for your employer. It encompasses construction, manufacturing and consultancy. In so doing you are making money from your engineering intellectual property. Those who hold this view argue that unless the skill or product can make a financial return there is little point pursuing it.

In reality it is a combination of these elements, and more. This was highlighted by Norman Haste, Chief Operating Officer for Laing O'Rourke, when he gave evidence to the Innovation, Universities, Science and Skills Select Committee (30 April 2008):

> '. . . engineers have to make business decisions; they have to understand where they interrelate with government; they have to understand where they interrelate with other branches of the engineering profession. I have floated the idea to many people that if I were designing an undergraduate course, I would include art, business management subjects, strategic planning, and possibly also something on politics.'[23]

Norman Haste is alluding to the reality that engineering is not practised in isolation; companies will expect their engineers to understand the business context in which they operate. For example there is little point in refining an engineering design or process if it does not improve the business return. Engineering is also sensitive to the financial and political landscape; sectors such as civil engineering and military products are particularly sensitive to current government policy because government departments are significant customers for those sectors. Norman Haste's comments fit with the findings of the Finniston Report[24] which identified that engineering education needed to be broader, incorporating subjects from business, languages and the arts. It was this report that ultimately led to the formation of the integrated MEng degree in the decade that followed.

Skills expected of all engineers

It is The UK Standard for Professional Engineering Competence (UK-SPEC)[21] published by the Engineering Council UK that informs the accreditation of degree courses by the Professional Engineering Institutions (PEIs) and thus the skills developed through university education. UK-SPEC sets out the range of competences that an engineer should acquire through their education, and each PEI will interpret these generic competences for their discipline and will also set expectations for engineers as they progress through their careers and gain greater experience. The competence and commitment statements within UK-SPEC for chartered engineers require that they can do the following.

- Use a combination of general and specialist engineering knowledge and understanding to optimise the application of existing and emerging technology. By using their underpinning knowledge of mathematics and engineering science the engineer can assess new materials and techniques for the creative and innovative development of engineering technology and a process of continuous improvement.

- Apply appropriate theoretical and practical methods to the analysis and solution of engineering problems. The engineer will follow a process of: conceive the opportunity – design a response – implement a solution – operate the result – and evaluate and learn from its performance. The design process is an important element of any engineering discipline demanding creativity and problem-solving skills. Problems may be technical or social. For example a construction site working at night in an urban setting was allowed to work provided the local residents were not disturbed. One response involved switching the warning buzzer on the crane from the standard horn to simulated bird song so that residents would find it less intrusive.

- Provide technical and commercial leadership. This includes the effective implementation of a project, including costs, scheduling, deliverables and team management.

- Demonstrate effective interpersonal skills in oral, written and graphical communications. The engineer must also demonstrate an ability to work productively with others, secure in a knowledge of

their own strengths and character and able to respond constructively
to the same in other members of a team.

- Demonstrate a personal commitment to professional standards,
  recognising obligations to society, the profession and the
  environment. This encompasses professional and legal codes of
  conduct, including health and safety legislation. Engineers are
  expected to adopt a high ethical standard and to take responsibility
  for their continuing personal and professional development (CPD).
  This heading also encompasses a responsibility to undertake
  engineering activities in a way that contributes to sustainable
  development.

Clearly engineering employers will expect a level of technical knowledge
and competence in the appropriate discipline for those entering a graduate
engineering role, but they also desire strong employability skills ('soft skills'). In fact
the CBI[4] found that employers rated employability skills as their highest priority,
followed by the specific degree subject.

The CBI[25] identified the following employability skills and attitudes sought by
graduate employers in general and surveyed the level of employer satisfaction (%
satisfied and very satisfied in 2012).

Application of information technology (94%)

Employers will expect a basic level of IT skills irrespective of the individual's
background. Engineers, however, will have a wealth of discipline specific IT skills that
they will have developed during their training, using software for activities such as:

- computer-aided design (CAD) including 3D modelling using AutoCAD,
  Solidworks, etc.

- systems and mathematical modelling using MATLAB, Dymola,
  Electronic Workbench, etc.

- computer programming languages

- specialist software specific to the discipline of study.

Sometimes an employer will use a package that is different to the one used during a student's studies, but it is important to recognise that the command and control structure of a software suite is just one part of the package; the more important and transferable part is the understanding of underlying principles and strategies in the use of that software.

Application of numeracy (90%)

Engineering education is highly mathematical and is one of the reasons why employers from other disciplines are keen to recruit engineering graduates. Engineering students should focus on their ability to analyse complicated numerical information and how to communicate it with clarity to others, less technically educated.

Communication and literacy (85%)

Depending on the nature of their work, engineers will liaise with agencies and individuals both inside and outside of their company. They must be able to present technical or project information to others in an appropriate level of detail and style for the audience, who might be clients, contractors, consultants, local and national government officers, residents or a supply chain company. Engineering students will find that they have experienced a wide range of communication forms in their degree course, such as laboratory and project reports, essays, posters, oral presentations, case studies, venture capital reports, web pages, etc.

Analysis skills (83%)

Engineers will use software and other techniques to ensure appropriate materials, section sizes, equipment, etc. are selected for the design and construction or manufacture of a product. As a top level competence embedded within the UK-SPEC, engineering students will be presented with ample opportunity to demonstrate this skill within their discipline specific curriculum.

Positive attitude to work (82%)

Both in employment and during their studies this attitude is easily demonstrated through one's attitude to work and especially when working as a member of a team.

Self-management (69%)

Self-management is concerned with how we organise our work and time in order to meet our commitments. Projects have timescales that must be met or they risk losing money for the company through overspend, loss of sales or penalties for late completion. Engineers tasked with bidding for contracts will be required to complete and submit those bids before the deadline set by the client; failure to do so will mean that any work undertaken to prepare the bid is wasted. Most students will be familiar with the experience of conflicting demands on their time and will have adopted appropriate strategies to ensure deadlines are met.

Problem-solving (77%)

Problem-solving is core to engineering, and subjects such as design and project work will develop the mind-set needed to successfully use the modelling and analytical tools developed in other modules.

Team-working (75%)

Very few engineers will work in isolation and most will be members of teams; in some companies these teams will have the added complication of spanning countries and time zones. Working constructively as a member of a team can be difficult where personalities, experience, objectives and approaches to work differ. For these reasons, this is one skill that many engineering students find causes them the greatest stress. Being a member of a team and placing one's future in the hands of others who may not share the same work ethic or schedule can be very difficult, as can finding that the team has a very capable individual who 'takes over' the task with little or no reference to others in the team. Good teamwork is about communication, honest debate and arriving at a distribution of work that allows 'the whole to become greater than the parts'. Some teams will contain a member who is a 'passenger' and contributing little to the effort and potentially creating more work for others. This problem is not unique to undergraduates and potential employers will be interested in how individuals and the team managed the situation (refer to Chapter 14 for more on this).

International cultural awareness (59%)

Engineers frequently work across boundaries.

- Some will work abroad for their company or will use the opportunity for mobility that their qualification affords. Here there may be challenges that arise from language or culture (see for example *My Story ... Steve Dobson* in Chapter 12). It is also possible that practices accepted in one culture are not acceptable in another.

- Those working in multinational teams using a common language also need to be aware of the potential misunderstandings that can arise across cultures. Such teams are increasingly common, both within and outside the UK.

- The need for cultural awareness is not restricted to different nationalities. Engineers may need to handle similar challenges across other boundaries such as those of ethnicity and company culture.

Universities are, by their nature, a mix of nationalities and there are ample opportunities for exploring this skill on the campus and thus demonstrating an awareness to potential employers. Other ways of developing and demonstrating international and cultural awareness are to undertake a project, work experience or study abroad as a part of a degree. Most universities will provide such opportunities and in some cases funding exists to support students undertaking international experience, an example being Erasmus funding (www.erasmusprogramme.com).

Business and customer awareness (53%)

Understanding how the engineer's work fits within the business and a commitment to customer care are valued skills for most employers. This was succinctly expressed by Hamid Mughal, Executive Vice President Manufacturing Engineering and Technology for Rolls-Royce.

'The danger with some graduate engineers is that they become too obsessed with detailed analysis and improvement of an element of production process – usually the technically appealing one – with perhaps no more than 0.05% impact on the total cost. In many instances, if they had taken a broader *view of the business they could well have put their technical effort in improving aspects of the total process chain that could yield significantly greater savings. At Rolls-Royce we need engineers who understand the importance of seeing the whole picture, not just one brush stroke.'[26]*

Many engineers will be involved in the preparation and administration of tender documents and contracts and liaising with their clients or customers. The extent to which engineering courses incorporate business content varies, but UK-SPEC demands that all courses include some business content. Specialist MSc degrees exist for those wishing to specialise in engineering business management.

Foreign language skills (46%)

This is less of a core skill for engineers than some other professions, but many universities provide opportunities for language modules and study abroad for those that wish to master a second language.

Many of the skills desired by employers can be developed during the degree course; however there are many other ways in which students can demonstrate some of these skills. The Wilson Review recognised that:

'Interactions with a community outside the university add further dimensions to self-confidence, experience and skills. Volunteering develops skills for education, employability and life and is therefore inextricably linked with generic skills development. University participation in networks of volunteering associations can facilitate student involvement with projects linked to local communities to underpin peer leadership, increase the social capital and the social mobility of students, and prepare students for employment through relevant links to local communities.'[27]

My story . . .

Alasdair Woodbridge, self-employed, business start-up Heat Genius Ltd

What did my engineering studies give me? . . . Answer: a master's degree and people who remain my closest friends 10 years on. Really, though, it was a set of skills and knowledge of how to apply myself, which have been and will continue to be useful for the rest of my life. I have always liked to try different things, as I feel what makes someone an interesting individual is the nature of the experiences they've had, rather than the depth of their knowledge in one particular field – a kind of jack-of-all trades and master of none.

For me engineering is all about maths and how one can use just a set of numbers to prove the most amazing things, from reassuring a client that the product will perform and be appropriate for their purpose, to perhaps demonstrating the potential performance of a component in a system. What can you do with engineering? Pretty much most things! My first engineering job was in an industry where I loved the end product and that was boats. I took a job working for a small yacht designer in Auckland, New Zealand as a CAD draughtsman. It was great to work for a one-man band as I saw all the aspects of the business. After 18 months it was time to return home and my next employer liked the fact that I had international experience. Three years later, having worked as a member of a small engineering team, I had learned a lot about how a small business works, and how a 50 metre sailing yacht develops from an idea in a meeting with a client to a product moored on a pontoon at the Monaco Yacht Show.

What did I like about my work? The people who I worked with, most of whom were similarly passionate about sailing and how stuff worked, an association and feeling of satisfaction regarding the finished product and the problem-solving aspects of taking a set of ideas and making them into reality. What did I not like about my work? Sitting in front of a computer screen 10 hours a day with little social interaction!

The sailing industry was hit, like many others, in the recession and I was made redundant. My thoughts turned to the industries of the future and settled on energy. We all know that the world is running out of oil and natural gas and no one, so far, has come up with a replacement fuel that is as transportable and versatile. A risk, but all my money (by which I mean my career) was now going to be focused on the energy industry and how engineering can get us out of a dependence on fossil fuels, and make the fuels we have left last longer. The only way I could get outside and not be tied to a desk, as I saw it, was to start my own company. No longer would I have to worry about someone telling me what to do or how to best spend my time. Great!

I had never worked in renewables or energy saving, I didn't have a list of clients I could take with me, or a portfolio of jobs that I had done in the past to show my expertise. All I had was a degree in engineering and a keen eye for a challenge. I researched my chosen industry and found that companies, government organisations and charities are only too keen to tell you all about how great their products, initiatives and ideas are. I had an idea that developed from what I learned about home heating systems and gas or oil boilers. Boiler design has improved massively over the past 10 years, but the bit that controls them has not. A 97% efficient condensing boiler is often being controlled by a 30-year-old bimetallic strip in a draughty corridor. Some houses don't even have a thermostat! Surely something better could be done using the technology we already have all around us?

Starting a business needs an idea and some grit and determination. I found a university friend with the skills that I lacked and two and a half years later our company Heat Genius is just about to be launched.

Postscript

This chapter has outlined a range of skills that are generic to all sectors of engineering. Where there are additional skills expected for individual sectors these will be discussed in the corresponding chapter of this book.

Part 2
The branches of engineering

5

Mechanical engineering

'Mechanical engineering is all about taking science and using it to produce things. It's about translating theoretical research into practical solutions and applications which are used by society.'[28]

Mechanical engineering is a diverse field of engineering and offers careers in a wide range of industries and this is reflected in the numbers of students admitted to mechanical engineering degrees each year exceeding 6,500. Mechanical engineers design not just the goods we buy and use, but also the machines and systems that were used to manufacture or process them before they became available to us.

The nature of the industry

Mechanical engineers find work in a variety of industries that need their design and analysis skills. They may work within manufacturing and construction companies or provide specialist design and consultancy services. As such the fortunes of mechanical engineering are closely tied to the civil and manufacturing industries, which are discussed in Chapters 6 and 12 of this book. The location for the manufacture of goods tends to be determined by cost and the availability of cheap labour, with the result that there have been numerous examples of companies moving their production to countries such as China and India. However, whilst there has been a loss of manufacturing volume to other countries, the skills and creativity of the mechanical engineer trained in the UK continue to retain their value and hence these aspects of engineering endure in the UK. Here are some examples of roles undertaken by mechanical engineers.

- **Asset management:** managing the whole life cycle (design, manufacture, commissioning, operation, maintenance, modification, decommissioning and disposal) of the physical assets of a company, for example maintenance of railway rolling stock.

- **Design of building services systems for construction projects:** such as heating, lighting, power, access and data networks.

- **Design optimisation and compromise:** modifying existing or new designs to optimise their performance or manufacture taking account of the complexity of interactions between competing factors such as weight, aerodynamic drag and stiffness, for example in a car. Full optimisation is not always possible and compromises must be made to make the product both functional and attractive to the potential customer.

- **Designing machinery:** for domestic and industrial use, taking account of its purpose and safety when considering ways in which it might be used or abused by being applied to tasks for which it was unintended. Simple examples include the design of safety cut-out devices to prevent injury by moving parts when protective covers around them are removed or opened. An example of misuse was demonstrated when an industrial vacuum cleaner, designed to clean up sawdust, was used to clean up waste aluminium powder; the powder reacted with the water used by the cleaner, creating hydrogen gas which exploded when the machine was next switched on.

- **Detailed design of pressure systems and mechanisms:** including stress and dynamic analysis of components and assemblies.

- **Management and business functions such as sales and marketing of technical products or services:** for example managing projects and teams of engineers. As engineers progress in their careers some will choose to focus on management rather than technical roles. Examples of engineering roles and career progression are discussed in Chapter 16.

- **Materials testing:** establishing the performance and/or quality of materials and components, for example testing for hidden cracks in safety critical items such as jet turbine blades.

- **Modelling automotive dynamics:** delivering the desired performance, comfort and safety specified for the vehicle.

- **Precision measurement:** whether it is time, mass, distance or volume, each has a financial impact if inaccurately measured. Precise measurement and control of surface finish, automated assembly using robots and neurosurgery all require precise measurement and positioning.

- **Prototype development:** developing, testing and evaluating theoretical designs using traditional (e.g. clay modelling of car bodies) or modern computer simulations and/or techniques such as rapid prototyping or 3D printing.

- **Research and development:** seeking to identify new products or processes and to improve on existing ones. This may also involve monitoring research literature and patent applications to identify developments that may be of interest to the company.

- **Sales engineer:** converting a client's technical needs into a specification and preparing quotations for the work. For example ITCM were responsible for designing the pyramid tea bag and packaging for Unilever. The company added value to the initial brief by looking beyond the conventional design of tea bags and working with the company to create a novel and better shape, creating space within the bag to promote better infusion of the tea. Their involvement ensured that the new design could be mass-produced. They designed an ultra-fast machine to manufacture the tea bags and also designed novel packaging for them. With the support of an associated marketing campaign, this engineered solution helped the brand to reclaim its position as market leader.[29]

- **Simulation of processes such as heat transfer and fluid flow:** for example modelling the effectiveness of solar panels, cooling of a nuclear reactor or air flow through or around a building.

- **Stress analysis of components:** using computer software to model the distribution of stress in components and thus anticipate any modes of failure and under what conditions they will occur.

Examples of employers of mechanical engineers include:

- Arup (www.arup.com): consultancy

- ConocoPhillips (www.conocophillips.co.uk): petrochemical

- EDF (www.edfenergy.com): energy generation

- Jaguar Land Rover (www.jaguarlandrover.com): automotive

- Mott MacDonald (www.mottmac.com): consultancy

- Perkins Engines (www.perkins.com): diesel engines

- Rolls-Royce (www.rolls-royce.com): aeronautical, maritime and energy

- Transport for London (www.tfl.gov.uk): transport services.

Main sectors

Mechanical engineers have roles in many engineering sectors including:

- aerospace

- automotive

- building services

- chemical

- energy

- manufacturing

- materials

- medicine and health.

For a discussion of aerospace, chemical, manufacturing and materials engineering please refer to Chapters 9, 8, 12 and 10 respectively.

Automotive

The UK has a history of car design and manufacture stretching back to the end of the 19th century. That history has been mixed and the casual observer might be forgiven for thinking that the industry was in terminal decline, but in fact the UK produces 1.5 million vehicles and 2.5 million engines each year, most of which are exported, constituting 11% of the UK's total exports. The automotive sector has an annual turnover of approximately £50 billion.[30] Brands with a presence in the UK include Aston Martin, BMW, Honda, Jaguar Land Rover, Nissan, Toyota and Vauxhall. There are also companies such as Caterpillar and JCB that produce vehicles for the construction industry. The UK is also a major centre for motorsport teams (e.g. Force India F1, McLaren F1, Mercedes AMG Petronas, Red Bull Racing and Williams F1) and support companies. There are well over 2,000 companies working in the automotive supply chain. The automotive sector is therefore a major employer; it is estimated that the sector employs 154,400 people in 3,220 workplaces across Great Britain, although not all of these are engineering roles.[31]

In an industry of this size, manufacturing a complex product, the roles for engineers are varied and may be focused on a relatively small part of the vehicle or be involved with teams tackling much larger issues. Here are some examples of the work undertaken by engineers.

- Designing and conducting testing programmes for the car and its components. This may involve physical testing but computer modelling is increasingly used.

- Designing maintenance programmes for the vehicles.

- Improving pedestrian safety in the event of an impact.

- Minimising emissions to meet stricter legislation and reduce the impact on the environment. There is significant interest in low carbon research and development within this sector.

- Monitoring the quality of production and of components provided by suppliers.

- Sourcing new materials and processes that are more sustainable.

Building services

The role of building service engineers covers a variety of services and features, but with a special emphasis on energy reduction in buildings. For large structures these can be complex systems that require careful design, installation, commissioning and maintenance. Engineers work with the following systems to ensure the desired environment is achieved:

- acoustics, for example in an auditorium or atrium

- air conditioning and refrigeration

- control, security and data systems, for example telephone, data networks and fire alarms

- energy and water, both from utility providers and captured from renewable sources such as solar panels, wind turbines and rainwater harvesting mounted on or near the structure

- escalators and lifts

- heating and ventilation

- lighting.

In some applications the engineer must ensure that the system is fail-safe, i.e. they must provide back up for critical functions, for example, a hospital must have alternative sources of electricity if the power is cut.

The size of this sector may be inferred from the 14,000 UK members of the Chartered Institution of Building Services Engineers. There is a general recognition that this sector is struggling to recruit sufficient engineers. The Construction Industry Council's Director of Skills was reported in 2008 as stating that:

> 'The current skills shortages will be heightened by the fact that 20% of most professionals across firms will be retiring in the next 20 years. Student numbers on built-environment courses have dropped 28% since 2003/04. The industry is struggling to attract high-calibre young professionals into the sector.'[32]

Energy

This sector overlaps with other branches of engineering is discussed in Chapter 8 'Chemical Engineering'. Fossil fuels, nuclear and renewable energy systems all need mechanical engineering expertise in the design of the systems (power stations, wind turbines, solar panels and the distribution networks associated with them) and modelling their performance in service. Examples of employers include:

- Atkins (www.atkinsglobal.com)

- EDF (www.edfenergy.com)

- Mott MacDonald (www.mottmac.com/markets/power)

- Siemens (www.energy.siemens.com).

Medicine and health

This sector spans a number of engineering disciplines, including pharmaceuticals (see Chapter 8 'Chemical Engineering') and biomedical engineering (see Chapter 10 'Materials Engineering'). There are parts of this sector that are easily identified as linking mechanical engineering with an understanding of the human body, including:

- artificial limbs

- the design and manufacture of joint replacements

- modelling physical diseases, for example syringomyelia in which fluid-filled cavities form in the spinal cord causing pain, muscle restriction and potentially lower-limb paralysis

- rehabilitation devices

- surgical devices including robotic and precision medical tools.

Employers can be found in research and development, manufacture, regulatory or healthcare settings, for example GE Healthcare (www3.gehealthcare.co.uk) and Siemens Medical Solutions (www.siemens.co.uk).

Entry routes

Accredited degrees in mechanical engineering provide the educational component for becoming a chartered engineer. Maths and physics are normally required subjects for entry to a degree course and further mathematics is a good third subject. Other routes into degree courses include HND and HNC courses or Foundation courses for those lacking the required entry subjects. Chapter 3 gives further details of entry routes to degree courses.

Graduate entry roles

Entry roles in larger companies will be shaped by the company's graduate training scheme, which will typically rotate the employee through different functions and departments in the company. The company EDF, for instance, runs a Nuclear Science and Engineering Graduate Scheme that is accredited by the Institution of Mechanical Engineers (IMechE) and is one year in duration. The scheme provides technical training together with courses on the business aspects of the company, including financial and commercial awareness. In contrast, Mott MacDonald provides a training scheme that is not fixed to a specific time period but is designed to provide flexible training supported by a professional mentor. The company's online training tool offers over 200 self-study courses that can be called upon as and when needed. This arrangement also meets the requirements of the IMechE's Monitored Professional Development Scheme (MPDS) as well as schemes for other professional engineering institutions. Chapter 16 discusses career routes for engineers, including MPDS and its equivalents in other PEIs.

Who will I work with?

Given the range of industries that recruit mechanical engineers a graduate may find themselves as part of a large mechanical team or in a multidisciplinary team working with other engineers such as chemical, civil, electrical, manufacturing, material and systems engineers. They will also interact with the commercial, technical and business functions according to the nature of their employer's business.

Salary levels

Graduate starting salaries average £25,700 and will vary according to size of company and location.[3] Chartered engineers will earn £35,000–£50,000 (www.

prospects.ac.uk) and a review of job advertisements reveals that salaries for experienced senior engineers are in the region of £60,000. Executives can earn significantly higher amounts. The mean salary is £42,500.[3] Some companies offer additional benefits.

My story . . .

Geoff Mayes, Project Manager for ITCM (www.molinsitcm.com)

Having graduated with a BEng Mechanical Engineering degree in 2006 I went to work for a company that manufactured aerospace components. My first job was as Project Engineer. The key elements of this role were to take an engineering drawing, determine how to manufacture the component or assembly, plan the work, progress it around the workshop and get it despatched. This covered most manufacturing processes as my company had machining, sheet metalwork welding, press work, heat treatment, inspection, including X-ray and dye pen testing, all on the one site. We would also regularly visit other suppliers to get more specialist work carried out. I progressed to the role of Project Manager and ran significant projects for both the Defence Evaluation and Research Agency (DERA) and BAe Systems. During this time I had a team of up to 30 people working under my supervision.

In 2003 I moved to ITCM, initially in a purchasing role but learning the design process in parallel. ITCM's position in the market place is to deliver high-speed packaging solutions that push the boundaries with respect to both speed and quality. Since joining ITCM I have been involved in many different projects for various different multinationals, GSK and Unilever to name but two. At ITCM my role has progressed from basic design and process development back to project management. However, where project management at my first employer covered purely the manufacture and assembly of components, the role within ITCM requires that I take responsibility for an entire project life cycle. This includes:

- putting together an initial scheme for a machine

- estimating the initial price for the project (anything between a few thousand and several million pounds)

- helping to win the job by assisting in the pitch to the client and estimating the price and time schedule for the work

- leading a team of mechanical, electrical and software engineers to design, build and develop a machine from a blank sheet of paper

- installing the machine

- training operators, etc. at the customer's site. ITCM has installed machines all around the world so this has required significant amounts of travel from time to time.

A typical project life cycle can take between one and two years with some extending to six or seven years. The rewards for the engineer lie in the sense of achievement you experience daily as you slowly make progress through the development of the machine and ultimately when you see a project that started as a blank sheet of paper come together to deliver and install a unique machine for the customer. Projects are frequently confidential, but here are some examples of the challenges that I have solved.

- Increasing a product manufacturing speed from 65 units per minute to 750 units per minute, achieving high line efficiency levels whilst making huge reductions in the number of operators required to run the lines.

- Delivery of a new high-speed weighing system to accurately check the quantity of pharmaceutical powder within a blister strip. This machine takes, for example, a 60 dose blister strip and automatically weighs each of the individual doses in six to eight minutes. In this particular case the machine process was relatively simple; however, whilst running the machine cycle care had to be taken not to affect any factors that would influence the weight reading. These include things such as air flow, magnetism, temperature, vibration, static electricity, moisture/humidity, machine level, pressure changes within the local vicinity, etc. Changes to any of these have a significant effect on the outcome of the test.

- Working with a partner company to deliver a new highly flexible, accurate, powder dosing system for patients. A significant challenge in this project is to achieve consistently accurate doses on a machine whilst maintaining flexibility. The aim was to avoid having to change parts when changing dose weights or powder type ranging from free flowing to very cohesive, sticky powders. The challenge includes feeding the powder, accurately measuring the doses and finally delivering the measured doses into the blister strip. Throughout the process a key restriction is that the machine and process must not change the powder properties in any way as this would render the product scrap.

6

Civil and structural engineering

'Civil Engineering is the practice of improving and maintaining the built and natural environment to enhance the quality of life for present and future generations'.[33]

'Structural engineers are specialists in design, construction, repair, conversion and conservation. They are concerned with all aspects of the structure and its stability.'[34]

Civil and structural engineering represent the earliest endeavours of mankind in engineering, with roots back to the earliest civilisations. Our lifestyle is dependent upon the work of civil and structural engineers, even more so as we move into a world where more than half of our population is living in cities. The discipline of civil engineering is very broad with a wide range of sub-disciplines and interest groups. Arguably structural engineering is one of these sub-groups, but given the structure of the industry and the representation of structural engineers within that industry it seems appropriate to link both civil and structural engineers together in this chapter.

We might consider the nature of civil engineering work in terms of the types of projects with which it is concerned, for example the supply of clean water and treatment of waste water. This includes the design of reservoirs, dams, distribution networks including pipelines, water treatment plants, water conservation and the control of pollution. Another themed approach is transport, incorporating the design of the road or rail network, signalling systems, the promotion of health and safety, congestion mitigation and maintenance. In the design of structures there is the consideration of the foundation design, structural skeleton, choice of materials, energy use and whole life costing, maintenance and decommissioning and providing the client with a structure or building that meets their requirements. It is apparent from these examples that within any type of civil or structural engineering project there will be several component stages or systems to be considered.

The nature of the industry

This industry is characterised by two main groups: consultants who are involved in design and contractors who undertake the construction works; however there are companies that undertake both roles, such as Morgan Sindall (www.morgansindall. com). There are also public service agencies and companies that employ civil and structural engineers. Examples of companies and agencies that fall under these headings are presented in Table 2.

TABLE 2: Examples of companies employing civil and structural engineers

Consultants	Contractors	Public services	Other companies
Aecom (www.aecom.com)	Balfour Beatty (www.balfourbeatty.com)	Environment Agency (www.environment-agency.gov.uk)	Network Rail (www.networkrail.co.uk)
Arup (www.arup.com)	BAM Nuttall (www.bamnuttall.co.uk)	Highways Agency (www.highways.gov.uk)	Severn Trent Water (www.stwater.co.uk)
Buro Happold (www.burohappold.com)	Galliford Try (www.gallifordtry.co.uk)	Local authorities	Transport for London (www.tfl.gov.uk)

The *New Civil Engineer* (NCE) publishes an annual assessment of the market for both consultants and contractors. For consultants the market is measured in terms of fees earned, whilst the contractor workload is measured according to turnover. The total fees for 2009–11 for the key sectors are shown below (the average total annual fees for the same period were approximately £13 billion):

- building: £7,991 million

- roads: £5,164 million

- rail: £3,691 million

- environment: £2,819 million

- water: £2,487 million.

Also explored in the NCE assessment is the level of contractor turnover. As demonstrated below, it too follows a similar pattern in demonstrating that the major sectors are transport (roads and rail), buildings and utilities (water and energy):

- roads: £2,890 million

- rail: £2,306 million

- water: £1,505 million

- energy: £694 million

- buildings: £446 million

- airports: £256 million

- flood defence: £229 million

- geotechnical: £162 million

- ports/harbours: £74 million

- defence: £53 million

- dams/reservoirs: £20 million.

Source: New Civil Engineer, August 2012

The nature of civil and structural engineering is such that successive governments have used the sector to both energise a slow economy by commissioning infrastructure projects and to dampen down an over-heated economy by restricting government expenditure. This does not mean that the industry is totally insulated from recession but it may limit the pain and in the boom times there is normally strong private sector demand in addition to essential government expenditure. In the UK the construction industry represents some 6%–10% of the Gross Domestic Product (GDP).

In contrast to most other branches of engineering, civil and structural engineering is characterised by the unique nature of the products caused either by the site conditions, e.g. geology or location, or because the client requires a structure that is distinctive. For civil and structural engineers the opportunity to be involved with the design and construction of a statement or iconic structure is one of the rewards of the job. Depending on the size of the project the period of involvement may be relatively short or many years in duration. Here are some examples of large-scale infrastructure projects in the UK and worldwide and their expected duration.

- **Crossrail, London** (2009–17): this latest addition to London's Underground network will run from east (Shenfield and Abbey Wood) to west (Maidenhead and Heathrow) through 21 km of new twin-bore tunnels.

- **UK Wind Farm Zones** (2009–20): this is an example of a civil engineering project within the renewable energy sector involving construction of offshore wind turbines located around the UK.

- **Akkuyu Nuclear Power Station, Turkey** (2013–19): this will have four power units each of 1200 MW and will generate approximately 35 billion kilowatt-hours per year.

- **Delhi Mumbai Industrial Corridor** (2008–16): this huge infrastructure project is an example of integrated planning and construction to provide support for industry and commerce with an eye to ease of export of the goods produced. The project stretches for 1,483 km between the political capital (Delhi) and the business capital of India (Mumbai). The project includes nine industrial zones, a high-speed freight line, three ports, six airports, a six-lane motorway connecting the two cities and a 4000 MW power plant.

- **Australian National Broadband Network** (2010–18): this ambitious utility project will provide 93% of Australian homes, schools and workplaces with optical fibre, superfast broadband connectivity. Wireless and satellite services will provide connectivity to the remaining 7% in remote locations.

- **Wuhan New Port, China** (2011–20): this is a significant expansion project to an existing strategically important port. The work will increase shipping capacity, cargo capacity and the associated dock side and transport infrastructure.

- **Zhuhai–Macau Bridge, Hong Kong** (2011–16): this sea crossing provides a three-lane carriageway via a bridge and tunnel link. It also includes the construction of two artificial islands.

- **Panama Canal Expansion** (2007–14): designed to increase the capacity of the canal, this project includes the construction of new sets of locks at each end of the canal. The works also include the widening and deepening of existing navigational channels.

- **USA High Speed Rail** (2011–22): this is a suite of 20 rail projects across the USA including new high-speed lines such as the one between Los Angeles and San Francisco.

- **World Trade Center, New York** (2006–15): this regeneration of the site destroyed by terrorist attack in 2001 includes five new skyscrapers, a memorial, museum, performing arts centre and retail space.

- **East Side Access Subway, New York** (2001–19): one of New York's transport infrastructure projects, this provides a new subway between Queens and Grand Central Station in Manhattan. It also includes the underground extension of Grand Central Station.

Main sectors

Civil and structural engineering work tends to be characterised by the particular nature or focus of the work and professional specialist interest groups exist to promote best practice. The main sectors are explained here and whilst many jobs will be advertised as civil engineer or structural engineer, others will be described in terms of the type of sector, e.g. graduate geotechnical engineer.

Geotechnical engineering

No civil structure is independent of the ground and it is the role of geotechnical engineers to establish the nature of the ground and to design appropriate works for the project. Examples of such projects are:

- foundations for buildings and other structures

- retaining structures to protect excavations or cuttings from collapse

- embankments and dams constructed from soil or rock

- design of slopes for excavations such as cuttings and quarries

- stabilisation of natural slopes.

Some geotechnical engineers specialise in establishing the site conditions through the process of site investigation or the more focused ground investigation. They may also be involved in testing the materials to establish their properties to allow for the subsequent design. With the increase in the reuse of sites (brownfield sites) that may contain pollutants or contaminated ground, the geotechnical engineer will be involved in designing remedial works to treat or isolate the contamination to ensure protection of the environment and future users of the site.

Highway and transportation engineering

The Institute of Highway Engineers (IHE) defines their role as 'Improving the highway environment to deliver a safe, sustainable transport system'.[35] Highway engineers design, construct and maintain the road infrastructure and this will include structures such as bridges, drainage, lighting, signage and support structures such as salt and grit stores.

Transportation engineers are concerned with schemes to provide balanced and sustainable transport solutions incorporating pedestrians, cyclists, public and private vehicles. They will design systems to promote safety, including traffic control signalling, and traffic calming measures, such as shared space schemes where pedestrians and vehicles have equality of presence.

Power and utilities engineering

Civil engineers working in this sector are supporting the distribution networks for electricity, gas, clean and waste water, and telephony services. To this end they may be involved in many of the roles described here such as tunnelling, water engineering, and the design and maintenance of structures. Engineers may work for the utility companies, specialist sub-contractors or government agencies. There are also specialist services that promote safe working in the vicinity of utilities, for example damage to a gas main can have very serious consequences.

Rail engineering

The UK's railways provide for well over one billion journeys each year and the engineers are responsible not just for the 20,000 miles of track, but also for over 40,000 structures such as bridges, tunnels, viaducts and stations. Railway civil engineers will therefore have a wide remit in terms of design, maintenance and refurbishment of the railway infrastructure. As well as the surface networks there are underground services, and London's underground is expanding and undergoing upgrading of a number of stations.

With the need to structure works around the operational timetables of the rail companies, the site work will often include nights, weekends and bank holiday periods.

Structural engineering

Structural engineers focus on the design and inspection of a wide range of building types from rollercoasters to skyscrapers, including:

- residential properties

- commercial and industrial units

- bridges

- theatres

- sports stadia

- hospitals

- schools, colleges and universities

- offshore structures such as oil/gas rigs and wind turbines.

They may work with clients and architects to determine the means by which their vision can be achieved and with contractors to deliver the structure. Their design will consider the performance of the structure as a whole, and in each part, for the forces and displacements exerted upon it, including dynamic forces created in its use or as a result of earthquake loading. Natural forces such as wind and snow loading must also be taken into consideration.

Increasingly, structural engineers are concerned with the sustainability of the structure, through the choice of materials and finishes, including the use of energy to produce, operate, maintain and decommission the structure.

Structural engineers will also inspect existing structures to ensure that they remain safe for use; examples include the assessment of bridges after being hit by a vehicle or exposed to a fire and the checking of buildings affected by earthquake, ground subsidence or gas explosion.

Tunnel engineering

Tunnels and underground excavations are common in urban settings and where there are natural obstacles such as water or mountains to be crossed. Underground spaces may contain car parks, shopping malls or other public amenities. In the UK a number of government infrastructure projects such as Crossrail, HS2 and tunnels for electrical distribution under London have major tunnelling components. With over 50% of the world's population living in cities and this figure set to grow, the pressure on city infrastructures will become greater and not just for transport, but also for provision of goods and utilities and removal of waste. Some visionaries imagine future cities as two systems: above ground will be the living and environmental systems for human health and well-being; and below ground will be the supporting transport and logistics systems.

The challenge for the tunnel engineer is to design the underground structures, access, excavation and support. In urban settings this will demand careful control and minimisation of ground movements and settlements around the tunnel to avoid damage to the existing infrastructure both on the surface and underground.

Water engineering

Water engineers work in a number of environments.

- **Coastal:** tackling issues such as coastal defences, waste water discharges and pollution of the inshore waters and beaches.

- **Clean water provision:** from source to use, the water engineer will be involved at each stage: accessing natural aquifers, designing reservoirs, dams, treatment plants, pumping stations, distribution channels and pipelines and the maintenance of the infrastructure to minimise losses. In some countries where water supply is limited, water engineers may be involved in the design, construction and operation of desalination plants or plants for recycling waste water for human consumption. In 2010 Thames Water opened the UK's first major desalination plant in East London for use in periods of drought. Much smaller plants have previously been installed in the UK, e.g. for the Millennium Dome.

- **Treatment and disposal of sewage and waste water:** ideally our waste water needs to be cleaned before it re-enters the water cycle. Cities like London face major challenges in this area because of old and sometimes inadequate systems that are still in use. In London's case this sometimes leads to discharge of untreated sewage into the Thames – 39 million tonnes of it each year. Engineers have proposed the Thames Tideway Tunnel to capture and store these overflows, allowing treatment and controlled discharge when the treatment plants have capacity.[36]

- **Flood control:** one in six properties in England and Wales are at risk of flooding and this risk is expected to increase. In Scotland one in 22 residential and one in 13 non-residential properties are at risk. The sources of flooding include:

  ○ rivers

  ○ coastal waters

  ○ raised groundwater levels

  ○ burst pipes, water mains and drains

  ○ hillside run-off after heavy storms.

  The devastation caused by flooding is costly to both the individuals affected and the economy. Water engineers will be involved in the design and construction of: flood alleviation schemes such as building embankments, flood walls, coastal defences, etc.; improved drainage systems; and the maintenance of the existing systems. During periods of flood risk, engineers may be called upon to work unsocial hours as they monitor and respond to the developing situation.

- **Pollution control:** the range of potential pollutants in our waterways is significant. Some are easily recognised but others are less obvious, for example chemicals used in our homes may be discharged into the waste water systems and find their way into our rivers. Other sources of pollution include run-off from roads in wet weather as the debris, dust and chemicals deposited on the road by traffic is washed into drains, catchment pools and waterways. Some pollutants will remain in the environment threatening the health and well-being of the natural environment including that of humans. The Environment Agency (www.environment-agency.gov.uk) and water utility companies such as Thames Water (www.thameswater.co.uk) and Scottish Water (www.scottishwater.co.uk) will be the main employers of engineers working on preventing and monitoring pollution.

Skills and values required

In addition to the generic engineering and employability skills outlined in Chapter 4 the following are typical of those needed to work successfully as a civil or structural engineer.

- Health and safety: promoting best practice from design, through construction to operation, maintenance or decommissioning. Engineers need to understand and conform to legislation such as the Construction Design and Management (CDM) regulations.

- Project management: monitoring projects or parts of projects for progress and budget and responding appropriately to deviations to the plan.

- Surveying and setting-out: may be a requirement for contractors, although specialist surveyors are often employed on site as required.

Entry routes

Accredited degrees in civil or structural engineering provide the educational component for becoming a chartered engineer. Maths and physics are normally required subjects for entry to a degree course and further mathematics and geography are good third subjects. Some universities combine civil engineering with architectural studies and will require evidence of artistic ability in the form of a GCSE in art or design or a portfolio. Other routes into degree courses are identified in Chapter 3.

Graduate engineers will find jobs advertised by contractors, consultants and others. Working for a contractor will help you develop skills and knowledge in a range of disciplines, whilst consultants offer the opportunity to develop specialist expertise and to work with a variety of clients and projects. Other employers may include those in the supply chain providing materials, water authorities, the Highways Agency, Environment Agency, Network Rail, and local and district councils.

Graduate entry roles

The type of work undertaken by new graduates depends on the company and the nature of their recruitment. Large companies may have formal graduate training schemes that rotate the individual through a range of departments or groups within the company to provide an insight into the nature of the company's work, values and structures. Smaller companies will be more focused and graduates may need to organise their own development plan. Most employers will want their recruits to work towards becoming a chartered engineer (see Chapter 16) and will have some form of training plan in place. Consultants and contractors may have agreements in place to provide experience of design or construction that the other cannot provide.

Graduates will work under supervision and may be given responsibility for small projects or aspects of a construction. As their experience grows they should expect to take on larger packages of work and more responsibility. It is in the nature of the work that civil and structural engineers will find themselves managing large budgets and potentially complex personnel structures.

With experience and especially if chartered, engineers will find ease of movement between companies and between consultant and contractor. A willingness to travel both in the UK and internationally is valued by many companies. Chartered engineers will find that their skills are recognised around the world. Other opportunities to gain international experience are to work with a non-governmental organisation such as:

- Engineers Without Borders (EWB) – for students and young professionals. EWB is an international development organisation that seeks to use engineering to remove barriers to development.

- RedR – a charity which provides skilled professionals to help save and rebuild the lives of people affected by natural and man-made disasters.

- WaterAid – transforms lives by improving access to safe water, hygiene and sanitation in the world's poorest communities.

Who will I work with?

Civil and structural engineers work in teams that draw together expertise from within their own industry but will also include others such as architects, lawyers, environmental engineers and clients. The size of the team will depend upon the nature of the project and may involve people working together from different centres, sometimes different countries. On site it is not unusual to find a range of sub-contractors working in their specialist areas and this requires careful project management, communication, and health and safety monitoring. The size of companies varies widely: in the 2012 NCE survey[37] the number of technical staff within a consultant company ranged from over 15,000 down to two, whilst the number of engineers employed by contractors ranged from almost 5,000 down to two. Figures 1 and 2 illustrate the distributions according to the NCE Survey and reveal that whilst there are a small number of large international companies there are many more smaller groups. This reflects the fact that for appropriately experienced engineers the start-up costs for establishing your own company are relatively low.

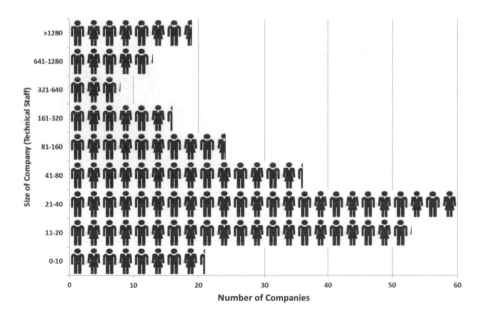

Figure 1: Size Distribution of Consultants (source data: 2012 *New Civil Engineer*'s Consultants File[37])

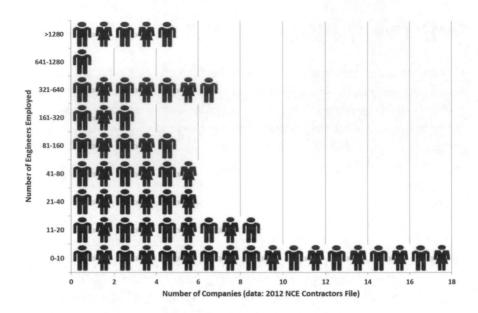

Figure 2: Size Distribution of Contractors by Number of Engineers (source data: 2012 *New Civil Engineer*'s Contractors File[37])

Salary levels

A sense of the current salary levels and progression can be gained by viewing one of the many recruitment websites. A typical starting salary for a graduate lies between £24,000 and £30,000 according to location. Experienced and/or chartered engineers will earn £30,000–£45,000 for senior engineer positions. Salaries in the private sector, for senior posts or in shortage areas, may be significantly higher.

My story . . .

Kate Cooksey, Senior Tunnel Design Engineer at Morgan Sindall Underground Professional Services (www.professionalservices. morgansindall.com)

- Whilst an undergraduate civil engineer I was sponsored by the Morgan Sindall Group and undertook vacation work and a one-year industrial placement working on the Scottish Water Quality and Standards project and two tunnelling projects, during which I worked on a number of tasks.

- Millport waste water treatment works: My role included construction of a new drainage facility, materials ordering, setting out and assisting with commissioning the waste water treatment works.

- Reservoir security: Preparing two week look-ahead programmes, method statements and risk assessments for the security upgrades to 1,500 reservoirs.

- Capital maintenance: Scoping and delivery of improvements to existing works in the interest of quality and safety. Writing and delivering health, safety and environmental documentation to ensure safe and sustainable control of works.

- Croydon cable tunnel (value £60 million): Design and construction of the site logistics for driving a 10km long tunnel and management of the site investigation at the drive shaft location.

- Kings Cross phase 2 underground station redevelopment (value £83 million): Checking engineer, verifying designs for the spray concrete tunnel lining.

After I graduated I was appointed Graduate Tunnel Design Engineer and I have worked my way up to Senior Tunnel Design Engineer. In the six years since I graduated I have worked on several high-value projects and have taken on more responsibility, including:

- a flood relief scheme (value £3 million)

- the Brixton to Honor Oak Thames Water Ring Main Extension (value £64 million)

- M74 extension (value £480 million)

- Lee Tunnel (value £435 million): a new sewer that is the deepest tunnel ever constructed under London

- preparing a tender for Crossrail for instrumentation and monitoring: I was responsible for scoping work worth over £100 million.

- Crossrail Whitechapel and Liverpool Street stations (value £235 million): as Instrumentation and Monitoring Manager for delivery of initial £45 million worth of work, I managed a team of 50 surveyors.

I have also been actively involved in a number of professional organisations, including the Midlands Branch of the Institution of Civil Engineers, being a committee member of the British Tunnelling Society (BTS) and establishing the BTS Young Members Committee which I chaired for the first two years. I was actively involved in establishing a Tunnelling MSc course at the University of Warwick and am a guest lecturer on some of the modules.

As a result of these activities I have been fortunate to win a number of awards:

- Women's Engineering Society Karen Burt Memorial Award Winner 2012 for best newly qualified female engineer

- *Management Today* 35 Women Under 35 List 2012

- BTS Young Members Committee winner of the *Tunnels and Tunnelling International* Sustainability Award: Investors in People, 2011

- Women of the Future Awards 2009, shortlisted finalist

- NCE Graduate Awards 2008, highly recommended finalist

I have really enjoyed the variety of work that I have been exposed to in my career, from the large multi-million pound projects through to the much smaller schemes. I learnt a lot from the breadth of work I have played a role in delivering and this all helped me to become a chartered civil engineer within five years of leaving university.

I get excited by being tasked with finding the solution to a challenging problem, however big or small. I also like the knowledge that the projects I am working on will help to improve the quality of life of at least one person. When I worked on a flood relief tunnel in Liverpool it was really nice to see what it meant to the local community. Meeting the residents and understanding that the solutions I'd provided would stop the need for sandbags and people to bucket water out of their neighbours' houses each year was a really great feeling.

I also find it exciting to see my ideas coming to life on site or being followed through by a team to make a way of working more efficient. When I designed the temporary works to connect a new drinking water tunnel to a large reservoir in south east London it was great to see the 1.4 metre diameter pipes being connected into the existing pipework maze and the final pieces of the network coming together.

I enjoy taking responsibility within a project for tasks and I strive, off the buzz of team energy, to find the best solution for a problem. I also really enjoy meeting, working with and learning from new people. It is always exciting to see what opportunities making new connections can bring. I have worked within many joint ventures due the nature of the big projects within the tunnelling industry and I find the new ideas and cultural variances between companies can really enhance what each of the partners contributes to the project.

There are so many engineering challenges out there to be solved to help make society a better place to live in and I am interested to see where my skills and those of the industry can lead, to help make an even better civilisation.

7

Electrical and electronic engineering

'Electrical engineering is a profession that uses science, technology, and problem-solving skills to design, construct, and maintain products, services, and information systems. Electrical engineering is the historical name for what is now called electrical, electronics, and computer engineering.'[38]

In its 2012 report the UK government noted that electronics is an enabling sector with the potential to 'offer widespread gains as the technologies they generate often drive innovation and productivity across the whole economy, or provide the solutions for end users that differentiate them in the market.'[39] The same report notes that past market failures have led to shortages of highly skilled electronics specialists.

The scope for graduate opportunities is indicated by the size of the UK electronics industry, which provides direct employment for over 300,000 people in 12,000 companies.[40]

The nature of the industry

In the Technology Strategy Board's (TSB) strategy for electronics, sensors and photonics[41] they identify that 8,000 companies in the UK's electronics industry generate £29 billion a year through design and manufacture. They anticipate significant opportunities for the sector in:

- plastic electronics

- power electronics

- electronic systems

- sensor systems

- photonics.

Electronics is identified by the TSB as an enabling technology for activities in 'healthcare, energy, transport, environmental sustainability, built environment and across the consumer market'.

Despite the promising future for electronic and electrical engineers and the demand for good graduates in a number of key industries, the numbers of students entering these degree programmes dropped by approximately 30% between 2001 and 2011. Electrical and electronic engineering is now the third largest engineering cohort after mechanical and civil engineering.[3] This decline occurred largely in the early part of the decade and may reflect the problems experienced by the industry at that time.

Employers of electrical and electronic engineers include:

- Alstom (www.alstom.com): power and transport

- CERAM (www.ceram.com): materials testing, analysis and consultancy

- GE (www.gepowerconversion.com): power conversion

- National Grid (www.nationalgrid.com): international electricity and gas company

- Nissan (careersatnissan.co.uk): automotive manufacturer

- Siemens (www.siemens.co.uk): energy, industry, infrastructure and cities and healthcare

- Thales (www.thalesgroup.com): aerospace, defence, security and transport.

Profile: Southampton Solent University

Study at Southampton Solent University: take the first step towards a rewarding career in engineering.

Our industry-focused approach

Based at the vibrant city-centre campus in Southampton, engineering at Solent is geared towards equipping you with hands-on experience from day one.

The university has a strong network of links with blue-chip companies – including Rolls-Royce, Stannah Stairlifts and computing firm CAPTEC – to ensure that students gain an in-depth insight into how they can transfer their skills to the workplace.

'Solent is unique in its approach – we combine lots of practical work and great facilities with a rigorous academic education,' says Rob Benham, course leader in engineering. 'We're now doing company visits for all first year students in the first three weeks and this industrial focus continues throughout the course.'

Live briefs and industrial projects are a regular feature, allowing students to put theory into practice and solve some real industrial problems – for example designing a conveyor belt attached to a robotic arm designed to enhance the production process in food packaging.

Our courses

You can work towards Chartered Engineer status on our two BEng (Hons) courses that have accreditation from the IET (Institution of Engineering and Technology) and the Engineering Council. A third BSc (Hons) course is new for 2013 and due to be reviewed for accreditation by the IET in the future. We also offer an HNC option which can lead to a degree.

- BEng (Hons) Electronic Engineering: accredited

- BEng (Hons) Mechanical Engineering: accredited

- BSc (Hons) Engineering Design and Manufacture: under review for accreditation

- HNC Engineering (part time).

A bonus of accreditation is the IET Solent network which hosts events at the university including guest lectures.

Our facilities
Solent has some of the best teaching and learning facilities in the country. The important practical element of our courses takes place in our recently refurbished laboratories with state-of-the-art specialist equipment and software, supported by a dedicated technician.

Key facilities include:

- two rapid prototype machines (Titan and Z-Corp)

- 2D laser cutter (Hobart)

- five-axis prototyping machine with 10' × 5' bed (Thermwood).

Our friendly academic support
A supportive learning environment is a priority at Solent. Tutors are always on hand to offer advice and encouragement both during group work – with small teams of students collaborating on experiments – and individual projects. They have years of experience in teaching and industry.

Our graduates
Solent graduates work in sectors as diverse as electrical, electronic design, communications, manufacturing, product design, mechanical and production engineering. And our industrial contacts prove invaluable after graduation, with recent students now employed at Stannah Stairlifts, Pascal Electronics, GE Aviation and Exxon Mobil.

Graduate John Ogden, who works at NATS (National Air Traffic Services), says: 'My time at Solent was a great experience – not only has it given me a good qualification in electronics, it has also helped me to achieve a great career as an air traffic control engineer.'

Your future in engineering
Contact us to take the first step.

Tel: +44 (0)23 8031 9975

Email: mt.admissions@solent.ac.uk

Case study

Martin Middleton, 26, was born and raised in Southampton and works for DP (Dubai Ports) World Southampton.

Juggling a degree with work pays off for Solent student

As a mature student, Martin studied part-time at Southampton Solent University on the BEng (Hons) Engineering with Business course. Martin graduated this year, winning first prize in the WRTI (Wessex Roundtable of Inventors) Technology and Innovation Awards for his final year project, a design for improving coolant recovery and reuse systems.

'Engineering inspires me. I enjoy problem-solving and the triumphs of technology such as large tunnels – it's a great sector to work in.

'So, studying for a degree to enhance my career prospects was the natural progression for me. I had achieved distinction grades for my HNC at college and, fortunately, my employers agreed to continue funding my studies at university.

'I chose Solent because it recognises the demand from part-time students for degree-level study and caters for their needs. I really enjoyed having access to all the latest technology and meeting like-minded people with similar interests. It's been difficult to work and study at the same time, but if you have any problems, the lecturers are all really approachable and always have time to speak even though they are busy.

'For my final year project I was aiming to improve the inefficient engine coolant procedures currently in place at DP World Southampton. I was totally surprised that I won the WRTI award as I genuinely thought the other projects looked better. It's given me new energy to drive the concept forward, fine tune it and see how far it can go.

'Throughout my studies, I have been lucky enough to have my company's support – they have paid for the course, released me on Wednesdays whenever I've been scheduled to work that day, and bought the materials needed for the prototype coolant system.

'I'd advise anyone thinking about studying engineering that it is such a big subject, covering many different sectors, that you have to make sure you know what you want to do and pick the right area. You will need a lot of enthusiasm and commitment, especially if you study part-time.

'For me the best thing about the degree is the feeling of accomplishment. I hope it will open up new doors and even lead to a job abroad.'

Precisely designed for industry

Solent's engineering courses deliver essential industry skills

With strong links to the industry and a well-established work placement programme, our engineering courses combine in-depth technical knowledge with the real, practical experience employers are searching for.

Some of the courses you can choose from:

BEng (Hons) Electronic Engineering

Get to grips with the full spectrum of electronic systems, from analogue and digital electronics to microwave comms and digital signal processing. Study real-time systems and microcomputers, and use the latest software to design solutions to industry-related problems.

BEng (Hons) Mechanical Engineering

Develop advanced skills in design, manufacture and product development, work on live briefs in collaboration with industry mentors, and get hands-on experience in our high-tech labs and workshops.

BSc (Hons) Engineering Design and Manufacture

Explore mechatronic systems and robotics, microcomputer systems and manufacturing principles in a professional engineering environment, and develop key business and entrepreneurial skills tailored to the industry.

Want to know more?

Visit **www.solent.ac.uk/courses**

Email **ask@solent.ac.uk**

Call **+44 (0)23 8031 9000**

Southampton
SOLENT
University

Main sectors

As an enabling technology electronics and electrical components and systems are found in a wide variety of products and sectors. The industry is composed of four broad customer bases:

- commercial applications: for example advertising screens, automated teller machines (ATMs) and credit cards with embedded electronics to select a method of payment between a rewards scheme and normal money transfer

- consumer applications: including TVs, radios, DVD players, smartphones, satellite navigation systems, refrigerators, etc.

- industrial applications: electric cars and trains, transport ticketing systems, building services control, water metering, etc.

- military applications: for weapons control, 'spy' satellites and space probes all with intrinsic higher-reliability and radiation hardened components.

Conventional electronics

Conventional semiconductor electronic systems for support of the above applications use circuit boards mounted with semi-conductor components such as diodes, transistors, integrated circuits together with switches, capacitors and resistors etc. Electronic engineers design the circuits to produce the desired output and will model them using pre-layout functional simulation and post-layout physical simulation prior to manufacture.

Biomedical engineering

Biomedical engineering involves the application of electronic and electrical engineering to activities such as:

- medical devices: such as heart pacemakers and cochlear implants

- medical imaging: for example magnetic resonance imaging (MRI) scanners

- physiological signal monitoring: of signals such as heart rhythm (ECG) and brain activity (EEG).

Computer engineering

This sector is concerned with the design of the hardware, software, communications and the interaction between them for computers and computer-based systems. Many products associated with the application fields discussed above require integration of hardware and software in embedded systems, that is, devices that have software and hardware embedded within them.

Control systems

Not all control systems require electrical or electronic components; for example the automatic regulation of water level in a tank is achieved with a simple mechanical device. However, the majority of control systems do. Here are some examples of electronic control system applications.

- Elevators in hotels that may have sophisticated algorithms for minimising energy, waiting time and/or electrical wear. These algorithms are implemented in embedded programmable electronic controllers that may use techniques found in Artificial Intelligence to be adaptive and self-learning.

- Smart buildings that use controllers for heating/cooling, lighting and ventilation.

- Industrial process control in, for example, a steel mill requires systems that will control the gauge, eccentricity, jumping and flatness of metal sheet production.

- Pharmaceutical manufacture and packaging requires accurate control of the individual dose and its encapsulation in blister packs etc.

- Automobile engine management using a powertrain control module to monitor multiple sensors and adjust the engine to run at optimum performance.

- Hard disk drives have sophisticated software-enabled electronic control systems for locating and following tracks.

Electrical generation and distribution

The UK electricity industry has three functions.

1. Generating companies which use a range of technologies from nuclear power through to small hydroelectric schemes. The generated electricity is fed into the national transmission network and on to regional networks.

2. Distributors who are the intermediaries managing the transfer of electricity between the generating and supply companies. They own and maintain the distribution infrastructure including the towers and cables that cross the country.

3. Supply companies who supply electricity to the consumer.

The 2007 European Commission 20-20-20 targets have set the agenda for, amongst others, the generating and supply companies, requiring a:

- 20% reduction in energy demand

- 20% reduction in greenhouse gas emissions

- 20% share of renewables in overall EU energy consumption

- 10% renewable energy component in transport fuel.

These targets combined with the challenges of an aging power infrastructure and a loss of traditional local fuel sources (coal) demand significant investment over the foreseeable future. An additional concern is how resilience can be built into the power generation and distribution system. Engineering the Future published a report in 2011 discussing the impact of climate change on infrastructure and observed that the electrical transmission system 'needs to be smarter and self-healing (switching automatically to different demand and power flows).'[42] Smart systems will better manage the balance between demand

and supply and will provide a driver for green technologies and distributed generation.

Embedded and cyber-physical systems

'Software enabled electronics embedded in product and processes are now increasingly the differentiator technology in both services and goods.'[39]

The use of embedded systems is arguably at a transitional stage that will revolutionise technology in much the same way as personal computers have. Historically, embedded systems were processors that were programmed to perform a particular task and they have been used in many applications. Indeed it is likely that you are within metres of several such systems for most of the time. You will find them in mobile phones, game consoles, domestic goods, cars, etc. The restricted nature of processors with a fixed purpose is changing as embedded systems begin to be used within cyber-physical systems (CPS). A CPS links the computational with the physical to create a networked system that is responsive to the physical world. CPS are typically closed-loop systems, where sensors make measurements of the physical world for processing in the cyber subsystems, which then respond through actuators that modify the physical world. The system is intelligent, being both reactive and predictive in its operation. Applications range from crash avoidance in cars to the futuristic Programmable Matter proposed by Campbell *et al.*[43] in which the response changes the nature of the physical matter itself. Electro-mechanical timers in domestic appliances have been replaced with more sophisticated microprocessor-based solid-state embedded programmable controllers. The former were often the purview of an electrical engineer while the latter are now that of the electronic engineer. If we imagine a communicating network of such controllers we have a cyber-physical system.

Wireless communications

Communication technologies include:

- short-range wireless communication such as ZigBee, Bluetooth and WiFi

- long-range terrestrial and deep-space communications of which telecommunications for audio and visual signals is a subset.

Between 1999 and 2009 the telecommunications industry has experienced incredible change, with the percentage of the world's population having a mobile phone subscription growing from 15% to 70%. There has been a consequential decline in public switched telephone network voice revenues.[44]

Photonics

A brief discussion of photonic materials is provided in Chapter 10; here we will focus on applications for photonics, which include:

- data transmission, storage, communication and networks

- energy efficient lighting

- laser systems

- medical and biological

- scanning, sensing and imaging systems

- screens and displays.

In March 2013 the UK government announced a £3.7 million investment to fund projects that apply innovative photonics technologies to solve challenges in the health sector, signalling their support of the sector's future.

Plastic electronics

This sector, also known as printed or organic electronics, has a global market that is forecast to grow to £25 billion by 2020 from £1 billion in 2012. The technology involves printing circuits onto almost any surface, including flexible sheets, replacing the need for conventional, discretely packaged silicon-wafer-based devices. Here are some examples of potential applications of this technology.[45]

- Addition of interactive elements to greetings cards and printed posters.[46]

- Tracking of manufactured goods and packaging through the printing of tracking devices on card- and plastic-based products.

- Incorporation of anti-counterfeiting elements in packaging.

- Embedding of an electronic clock function allowing, for example, pharmaceutical companies to manufacture drug packaging that advises the user when the next dose should be taken. Similar systems on packaging might also monitor the shelf-life of products.[47]

- Addition of sensor systems within civil engineering structures for monitoring their behaviour.

- Manufacture of building facia and roofing sheets with integral solar photovoltaic modules.[48]

Power electronics

Power electronics is the application of electronics to the efficient management of power from milliwatts to gigawatts. It has applications across a wide range of industries and size of product, including mobile phones and electric-hybrid vehicles. Power savings of up to 40% are achieved on conventional electrical devices and this is expected to grow. In 2011 the UK government's Power Electronics Strategy Group expressed their belief that 'Power electronics is driving significant change in almost all sectors of industry, but identified particularly exciting opportunities in aerospace, automotive, renewable energy generation, energy networks and industrial processes.'[40]

The same group estimated that in 2011 power electronics had a direct global market worth £135 billion and was growing at an annual rate of 10%. This growth will drive a demand for engineers with the relevant skills.

Entry routes

Access to graduate careers in the electrical and electronic industries requires an undergraduate degree in electrical and/or electronic engineering or an aligned discipline such as:

- automotive electronic engineering

- communications engineering

- digital systems engineering

- electrical power engineering

- medical electronics and instrumentation

- microelectronics

- physics and photonics

- telecommunications engineering.

Graduate entry roles

According to a report produced jointly by the Higher Education Careers Service Unit and the Association of Graduate Careers Advisory Services,[49] electrical and electronic engineering graduates were employed according to the following distribution:

- 36.2% as engineering professionals

- 19.1% as information technology professionals

- 6.1% as commercial, industrial and public sector managers

- 3.6% as other professionals, associate professional and technical occupations

- 35.0% in other roles.

The recruiters will range from banks that need people to ensure their systems function and are secure through to automotive and aerospace companies. Many companies will recruit from a range of graduate disciplines to service their various functions and therefore the graduate training schemes for electrical and electronic graduates will be similar in structure to others, typically offering a rotation through different departments within the company.

For those entering electronics and electrical engineering typical roles include the following.

- Asset management: as well as managing existing assets, identifying needs for new assets or decommissioning them. For example National Grid[50] must manage their substations, cables, overhead lines and compressor stations. Responsibilities include maintenance, fault repairs and managing national contracts for activities such as pylon painting and vegetation control.

- Cost analysis to ensure optimum profitability.

- Design and application of automotive electronics: for example audio and IT instruments, switch systems, wiring harness and power systems.

- Design and layout of printed circuit boards.

- Developing electrical power generation and distribution systems across a range of application scales, e.g. cars, ships, national.

- Developing models and prototypes of products using three-dimensional design software.

- Developing next generation electronic and IT systems: for example satellite navigation or mobile phones.

- Drawing up and presenting plans and specifications for projects using software.

- Electronic and hardware design of products: including component selection, simulation, integration and schematic capture (such as computer aided design and electronic design automation systems).

- Ensuring compliance with statutory regulations.

- Interpreting and realising technical requirements.

- System level design: for example electric trains and signalling; weapons telemetry, guidance and control; and marine propulsion with electric motors.

- Translating customer requirements into high-quality, reliable demonstrators, prototypes and products.

- Validating all aspects of electrical and electronic systems: including environmental and EMC testing.

- Writing embedded programmes: examples include in-car navigation and combined information and entertainment (infotainment) systems and battery chargers for portable devices.

- Writing technical documentation: including change records, design technical notes, test specifications and test results.

Salary levels

The median graduate starting salary predicted by the Association of Graduate Recruiters in 2012 was £25,500.[3] The Prospects graduate careers website[51] reports starting salaries in the range £18,000–£29,000 increasing to £35,000–£45,000 for chartered engineers. Senior engineers earn £40,000–£55,000 and opportunities exist for much higher salaries. The average salary for all electronics engineers is £45,558 and £44,898 for electrical engineers.[3]

My story . . .
Ed Hall, Electrical Engineer at Siemens Wind Power

I successfully applied to the Siemens Wind Power Graduate Scheme in August 2009, pretty much directly after securing my MEng degree. The role focused on operation and maintenance activities on Siemens-built wind farms in the UK and Ireland, both on and offshore. As with many engineering based graduate schemes, it involved a rotation around a number of areas of the organisation to gain maximum exposure and understanding of how the business works. This included 'hands on' time focused on the turbines as part of site maintenance and servicing activities, allowing an in-depth knowledge of the product to develop. I was also involved in meetings and discussions with Siemens' customers, allowing my commercial awareness to progress.

In August 2011 I progressed from the scheme into a permanent role as an Electrical Engineer. Essentially, the role has responsibility for the technical integrity and performance of all electrical components on Siemens wind turbines operating within the region. This includes activities such as:

- performing analysis on components, both in the field and remotely

- completing investigation reports on defective components

- developing long term and effective solutions to remedy defective components

- working with suppliers and customers to maximise the productivity of the turbines.

All this is done through a strong working relationship with our technical headquarters in Denmark.

The fact that I'm deeply involved in the development of the UK's renewable energy industry is something that really motivates and excites me about my daily work. The change in the UK's energy supply in order to adopt significantly more renewable sources in the last few years is something that is now being taken really seriously not just in the technical world, but throughout society, and to be so involved in the day-to-day development of this is very rewarding for me.

Additionally, the fact that the industry is relatively young means that there are always areas that can be further developed, so it's a great opportunity to really make your mark. It also means that, for me, no day at work is ever quite the same. There are always new challenges to adapt to.

8

Chemical engineering

'Chemical engineering is all about identifying a safe and cost effective way of changing raw materials into useful products that you use every day.'[52]

Chemical engineering is distinct from other engineering disciplines because it is not about assembling components but rather about changing the state (or phase) of a compound. Chemical engineers are required to take a holistic approach to the problem and must understand both the individual processing steps, called **unit operations**, as well as the project as a whole. Examples of unit operations include reactions, distillation and separation. Chemical engineers understand both the chemistry of the processes that they use and the engineering components of the systems that are constructed to process the raw materials, for example distillation columns, heat exchangers and chemical reactors. They will be concerned with optimising the process to provide maximum yield or quality of product in an energy efficient and safe manner. They will also be concerned to minimise waste by identifying ways in which by-products of the process can be utilised. Chemical engineers are sometimes described as process engineers when their work is focused on designing or managing a process.

Calling all chemical engineers!

Don't trip up your career before it gets going – update your IChemE membership to ensure the best support on your route to becoming Chartered. With Chartered Chemical Engineers earning £10k a year more* – it pays to start your journey now!

Not a member? Joining is easy – simply use the QR code below or visit us at **www.icheme.org**

Looking for a job? IChemE is committed to the professional development of its members. Visit our careers pages to find out which companies offer an IChemE accredited company training scheme (ACTS) at

www.icheme.org/acts

*IChemE's 2012 salary survey shows that Chartered Chemical Engineers in their mid-30s earn on average £10k more than their non-qualified peers.

IChemE
ADVANCING
CHEMICAL
ENGINEERING
WORLDWIDE

0721_13

Profile: Institution of Chemical Engineers

You know the subject, you have the degree, and now you want to find that perfect job.

Your career search will pose many questions: Who do I work for? How do I ensure a good start to my career? What is my next goal?

So how do you go about making these all important career choices? This is where your professional body comes in – IChemE provides a benchmark to help you understand your career options and is here to support you as you move forward in your professional development.

Get your ACTS together!

As graduates entering this dynamic industry, it is vital that you continue to develop your knowledge and skills, and seek opportunities that contribute towards your professional development. So why not apply to a company that puts your development at the centre of their business? Organisations with an IChemE Accredited Company Training Scheme (ACTS) offer well-structured programmes, giving you the opportunities you need to develop your skills and gain the experience that will help you become a competent professional in the shortest time frame. Plus you will benefit from the experience of a Chartered Member of IChemE as your mentor to support your development. All of this combined fast-tracks you towards achieving the gold standard qualification of Chartered Chemical Engineer MIChemE.

'Working at KBR has really helped me progress towards achieving my goal of chartership by offering a diverse and interesting range of challenges whilst supporting and advising me through their well-structured mentoring scheme.'

Marcus Southern, KBR (UK)

Find out more and contact companies direct: www.icheme.org/acts

On top of this, some employers go beyond ACTS to prove their investment throughout your career as a process engineer. At IChemE we recognise these best players in industry as Corporate Partners.

The nature of the industry

Chemical engineers are often associated with large processing plants for oil or gas, but in reality the industry is much more diverse than this. They are also involved in mining, materials processing, nuclear and renewable energy systems. The range of industries that depend upon the processing of chemicals and materials places the chemical engineer in great demand. Example applications include the production of:

- agrochemicals

- chemicals

- clean water

- fibres

- food and drink

- metals

- paper

- personal care products

- pharmaceuticals

- plastics

- polymers.

The industry can be characterised by three groups of companies.

1. Operators who process the raw materials to provide finished products.

2. Contractors who provide professional services to the operators, for example the design and construction of processing plants such as refineries and pharmaceutical manufacturing systems.

3. Equipment manufacturers who design and manufacture the components of a processing plant.

Examples of each are given in Table 3.

TABLE 3: Examples of companies employing chemical engineers

Operators	Contractors	Equipment manufacturers
BP (www.bp.com) *Oil, gas & alternative energies*	Bouygues Energies & Services (www.bouygues-es.co.uk)	ITCM (www.molinsitcm.com)
GlaxoSmithKline (www.gsk.com) *Healthcare*	Foster Wheeler Energy (www.fwc.com)	Larsen & Toubro Ltd (www.larsentoubro.com)
Monarch Chemicals (www.monarchchemicals.co.uk) *Chemicals*	Jacobs Engineering Group (www.jacobs.com)	Marchant Schmidt (www.marchantschmidt.com)
Tate & Lyle (www.tateandlyle.com) *Food & drink ingredients*	WSP CEL (www.wspcel.com)	Poole Process Equipment Ltd (www.poole-process.co.uk)

Main sectors

As we have seen above, the industries within which chemical engineers work are extremely diverse, reflecting the range of goods produced for domestic, agricultural and industrial use. The roles of chemical engineers can be categorised under the following sectors.

Biochemical engineering

This sector focuses on the design of processes that involve the use of biological organisms or molecules, for example a bioreactor may be used to produce hydrogen from algae. Other techniques include separation using chromatography, filtration and centrifuges. As with chemical engineering in general, biochemical engineering is concerned with taking laboratory processes and scaling them up to commercial production by identifying the optimal conditions for cell growth, protein production or protein and virus expression and recovery.

Profile: University College London

A future in biochemical engineering

Biochemical engineers translate exciting discoveries in life sciences into practical materials and processes contributing to human health, sustainability and a cleaner environment. If you are interested in applying your skills and knowledge to meet global challenges relating to the development of novel medicines, pioneering stem cell therapies or green sustainable technologies, then biochemical engineering is for you. Below we describe two of the different sectors you could be involved in:

Medicines for the 21st century: therapeutic proteins

Over the past few decades, medical care has been revolutionised by the introduction of a whole class of drugs – therapeutic proteins. These large macro-molecules occur naturally in our bodies but only in very minute quantities. To be used for therapy we have to devise efficient means to generate much larger amounts, to ensure the product is incredibly pure and then supply it to the patient as a therapeutic drug, e.g. insulin for diabetes or antibodies to combat cancer.

Creating a process for making such a therapeutic protein can take many years of careful development and testing. So whilst data is being generated to prove the clinical use of the drug, the biochemical engineers are working flat out to ensure that the molecule can be made at large scale, consistently and of course at a price we can all afford! Development costs can be huge, exceeding $1 billion!

Green chemistry

As scientists and engineers we have a crucial and challenging future role in reducing the impact that our manufacturing processes have on our planet and its climate. While many claim that governments should do more, those of us with a scientific education are probably the only ones who can change the way industry operates. Indeed we also have a responsibility to do so.

Much of industrial chemistry has been carried out for the last century without regard to sustainability or environmental principles. Energy and raw material consumption, and chemical emissions, were utilised at whatever levels enabled the job to be done as quickly and cheaply as possible.

However, this view is now changing very rapidly with new guidelines aimed at creating clean, 'green' processes with minimal energy consumption and reduced waste. One of the most exciting and innovative ways that this can be achieved is to use nature's catalysts to perform chemical reactions more efficiently, and most importantly, under mild conditions which have minimal environmental burden. Nature's biocatalysts are either enzymes or even entire cells. The study of these biocatalysts, and the development of processes which use them to create new biomaterials and therapeutic reagents, are a cornerstone of the rapidly expanding discipline of biochemical engineering.

Case study

Career options

According to national statistics, employment prospects for biochemical engineers are excellent. Biochemical engineers are sought by all sectors of the biotechnology industries, including those with interests in pharmaceuticals, food, environment, consulting, contracting, management and finance. New graduate biochemical engineers are able to fulfil their individual aspirations and help to create safe and environmentally acceptable products. Their vital contribution to the national wealth helps to create the healthy economy that underpins a caring community.

The UK biotechnology industry is the leader in Europe, and second only to the USA, a fact acknowledged by the level of support from the UK government's Department for Business, Enterprise and Regulatory Reform (BERR). The biotechnology industry has an excellent record of investment, to ensure it stays at the cutting edge. This is not only in equipment and facilities but also in the people it employs. UCL biochemical engineers are taught to think systematically and to calculate precisely. With such large amounts of money at stake in the creation of each new medicine, well-informed investment decisions in the industry are a crucial responsibility.

Biochemical engineering at UCL

UCL pioneered the teaching of biochemical engineering in the UK and has been providing the process industries with first-class people ever since it started teaching the subject in 1956. The department now has a unique national and international network of industrial contacts which help it to maintain its position at the forefront of education and research. It is the largest centre in the UK and was ranked as 'internationally leading' in the last national Research Assessment Exercise. Undergraduates, masters and research students are taught first-hand about the most recent breakthroughs as part of their studies. UCL Biochemical Engineering provides a world-class teaching and learning environment for students to thrive and acquire these creative skills as well as understand the ethical and social impacts.

Bioproduct engineering

Bioproduct engineers develop sustainable solutions to meet the world's needs, such as biofuels and biodegradable plastics. They are also concerned with the use of processed natural and sustainable materials such as biomass. The term 'biomass' describes agricultural products either grown and harvested as a renewable source or produced as a by-product of other manufacturing activities such as waste generated in timber or paper mills. Bioproduct engineering uses biomass feedstock to manufacture chemicals, fuels for transportation or to generate electricity.

Energy

The traditional energy sources of oil, gas and coal have a limited life as non-renewables and also present challenges in terms of controlling the pollution associated with their extraction and subsequent use. Nevertheless, our dependence on fossil fuels remains significant and on a global level we are a long way from achieving a sustainable energy supply. The World Energy Outlook 2012 published by the International Energy Agency (IEA)[53] reported:

- CO_2 emissions at a record high

- a reduced energy efficiency of the global economy

- spending on oil imports at near-record levels.

The BP Energy Outlook 2030[54] anticipates a 39% increase in global energy consumption between 2010 and 2030 due to the rapid growth of low and medium income economies. To meet this demand, the contribution of renewable sources, including biomass and hydroelectricity, are predicted to grow, as is nuclear energy. However, BP expects the use of conventional carbon fuels (coal, gas and oil) to remain the major component of our energy economy in 2030. Chemical engineers are playing an important role in developing and using renewable fuels for transportation and electricity generation. They also work to ensure safe processes within the nuclear energy sector.

Renewable technologies also include harnessing solar and wind energy. Chemical engineers are trained in the development and use of materials as well as heat transfer and are therefore well placed to develop the technologies needed for efficient energy conversion or the design of materials and sensors for use in wind turbines.

Food processing

The manufacture of food products for human and animal consumption involves a range of processes that must be carefully designed and maintained. Consider for example the processes required to produce a dog food in regulated quantities that ensure the publicised weight is delivered. The raw ingredients will first need to be weighed and mixed before being placed in the final clean can, tray or pouch. The sealed containers are then cooked, cooled and labelled ready for distribution. The chemical engineer will be concerned with each step in the sequence and the means by which the materials are transported between stages in such a way as to avoid contamination.

Oil and gas refining

Crude oil and natural gas are processed in refineries to create the fuels and numerous products used for everything from the tarmac on our roads to the synthetic fibres used to manufacture the clothes on our backs. Refineries are large, complicated systems with safety demands that require chemical engineers with skills in process design, sensors and systems for monitoring the process, and risk assessment. Their responsibilities will include oversight of maintenance and the planning and execution of the safe shutdown of equipment and processes to allow this.

Pharmaceutical

Both pharmaceutical research and production facilities require careful design of the manufacturing process to meet stringent quality, health and safety requirements. From the initial drug development and trials, the engineers will find ways to scale up the procedures to achieve full commercial production levels. Pharmaceutical products are delivered in different forms such as tablets, powders, capsules, sprays and liquids and each must be manufactured in precise quantities, packaged and dispensed appropriately. Engineers will work to achieve this efficiently without risk of contamination to the product, uneven distribution of active ingredients or escape of the drug to the manufacturing environment with the risk to employees working there.

Water engineering

Our society uses and disposes of significant quantities of waste water that needs to be treated to protect the natural waterways and/or to facilitate its reuse. The design of sewage treatment works and processing plants for water extracted for domestic use requires engineers with an understanding of the chemical processes required. In some countries, such as Singapore, used waste water is reclaimed and processed using advanced membrane technologies and ultra-violet disinfection to deliver purified drinking water. Desalination plants use reverse-osmosis to convert salt water into drinking water. Many industrial processes use significant quantities of water that must be cleaned of pollutants before being released into the environment. Chemical engineers also work to design processes and products that prevent the release of potentially harmful chemicals into the water supply through their use, for example products used in domestic cleaning such as soap, washing liquids/powders and toilet cleaners.

Entry routes

The route to a career in chemical engineering is normally via an accredited degree course. Maths and chemistry at A level are expected subjects and physics or biology are good third subjects. Other routes into degree courses are identified in Chapter 3.

Graduate entry roles

A survey of the employers advertising in the Graduate Supplement of the September 2012 publication *The Chemical Engineer Today* reveals that many large companies across a diverse range of industries planned to recruit graduate chemical engineers in 2013. Some of these graduates will be entering specialist groups within larger companies selling consultant or professional services to the process industries, whilst others will work for producers of goods. Many graduates will work in a smaller group of chemical engineers with the advantage that they will have the opportunity to gain a breadth of experience early in their career. In a larger organisation employing many chemical engineers, such as the oil and gas industry, graduates will work under supervision within a team setting and be engaged in projects such as scoping a problem, reviewing the performance or potential for technologies, making cost estimates for maintenance, purchase or installation of equipment, designing new systems or commissioning new equipment or processes. The nature of their work may mean that they are involved in decisions with expenditure implications running to millions of pounds.

Who will I work with?

Because chemical engineers are distributed across so many industries and applications they will often form part of a larger interdisciplinary team working with other engineers such as civil, electrical, manufacturing, material, mechanical and systems engineers. They will also interact with the commercial, technical and business functions according to the nature of their employer's business. In some companies the chemical engineer may be the sole expert and must therefore apply their skills to a wide variety of projects or tasks.

Salary levels

According to the Institution of Chemical Engineers' Salary Survey 2012,[55] the median annual salaries in the UK and Ireland in 2011 were £27,861 for new graduates, £53,000 across the sector and £71,000 for chartered engineers who had completed a first degree. The oil and gas exploration and production sector enjoyed the highest median salaries of those working in an engineering function.

9

Aeronautical and aerospace engineering

*'Aerospace engineering is largely about artefacts that fly –
airplanes, rockets, satellites, missiles, etc.'*[56]

Aeronautical engineering is concerned with the design, construction, propulsion, control and safe operation of aircraft, whilst aerospace engineering extends the scope of aeronautical engineering to include spacecraft and satellites. Therefore when this chapter uses the term aerospace it refers also to aeronautical engineering.

Aerospace engineering includes topics such as:

- aerodynamics: the study of air flow over the structure to maximise performance

- aircraft dynamics: modelling how the aircraft behaves in the air

- control theory: the mathematical techniques for monitoring and controlling a system such as the propulsion or flight controls

- fluid dynamics: the underpinning science of fluid flow over surfaces and through pipes

- manufacturing processes: how aircraft space frames, engines, etc. can be manufactured to achieve the necessary precision and quality

- materials: working with a wide range of materials, many of which will be used in extreme conditions

- simulation: using computers to predict the behaviour of components, sub-assemblies and aircraft

- stress analysis: analysing the stress distribution within components and structures to identify locations of potential failure and to optimise design

- structural design: of the aircraft body and wings.

The nature of the industry

Aerospace manufacturing produces advanced, high-value-added products, a point vividly illustrated by Sir John Rose, Chief Executive of Rolls-Royce, when he said, 'Pound for pound an aircraft engine is six times more valuable than silver, whereas pound for pound, a motor car has the same value as a hamburger.'[5]

For both this reason and the fact that the UK aerospace industry has a 17% share of the global market and is second in size only to the USA, the UK government has committed to investing in the industry. The government has noted the opportunity provided by the projected growth in worldwide demand for aeronautical products by 2030 including:

- 27,000 new passenger aircraft, value c. £2.5 trillion

- 40,000 new commercial helicopters, value c. £109 billion.[57]

As a result, the UK aerospace industry is expected to grow at a rate of 6.8% over the next few years and the Confederation of British Industry (CBI) have stated that if the UK maintains its current market share, air traffic growth in Asia alone could generate 20,000 high-value jobs in the UK aerospace industry in the next 10 years.[58]

Although the UK still produces helicopters and military vehicles, its position in the global market place has been characterised by a move away from designing and manufacturing complete aircraft to concentrating on components and subsystems.[59] Aerospace companies within the UK are leaders in the design and manufacture of:

- aerostructures: the components that make up the structure of the plane, e.g. the fuselage

- aircraft systems: examples include electronic systems to navigate or control the flight of the plane, sometimes called 'fly by wire'.

- engines: for example a Rolls-Royce Trent 1000 engine has 66 turbine blades that operate at temperatures of up to 1,600 degrees centigrade. Each blade is grown from a single crystal of a specialist alloy and has dimensions accurate to 10 microns (one-tenth the diameter of a human hair)

- wings: for example the Airbus A380 wings are manufactured in North Wales and transported to Toulouse in France where the plane is assembled (the reader may be interested to reflect on the engineering and logistic challenges posed by transporting a wing that is 50 metres long, 14 metres wide and weighs 36 tonnes by river, sea and road between the two manufacturing plants)

- helicopters.

Because of the high-value-added nature of its products, the aerospace industry invests significantly in research and development (R&D). Examples of employers of aerospace engineers include:

- Airbus (www.airbus.com): aircraft manufacture

- AgustaWestland (www.agustawestland.com): helicopter manufacture

- Astrium (www.astrium.eads.net): space including launch and satellite systems

- BAE Systems (www.baesystems.com): military aircraft and systems

- J2 Aircraft Dynamics (www.j2aircraft.com): aircraft design and simulation software

- Rolls-Royce (www.rolls-royce.com): jet engines.

Within the UK, the aerospace industry supports more than 3,000 companies and directly employs 100,000 people, approximately 7% of whom work abroad.[60, 61] A reported shortage of graduate engineers with the appropriate skills has led to the industry and UK government co-funding bursaries for up to 500 aeronautical engineering related MSc course places over three years, commencing in 2013. Details of the scheme's requirements, eligible courses and supporting companies can be found on the Royal Academy of Engineering's web pages at www.raeng.org.uk/education/aeromsc.

Profile: Royal Aeronautical Society

Want to work in one of the world's most exciting industries? Become a member of the Royal Aeronautical Society and we'll help you reach your potential

Who we are

The Royal Aeronautical Society (RAeS) is the world's only professional body dedicated to the entire aerospace community. We exist to further the art, science and engineering of aeronautics by promoting the highest professional standards across all aerospace disciplines.

As a multidisciplinary organisation, the Society's membership profile is drawn from a diverse range of organisations and fields of expertise, including engineering, design, aircrew and air traffic control, along with professions that serve the aerospace, aviation and space industries, such as law, finance, marketing and recruitment.

We offer **free** membership to those studying full time and additional grades of membership depending on your experience and the stage you are at in your career. Some of our benefits include the following:

Free careers advice and guidance

Our dedicated and knowledgeable team will support you in your career development and offer one-to-one advice and guidance such as CV and covering letter writing.

The Society also organises a careers fair every autumn where companies across the industry recruit many talented engineers and we have an online career centre where you can browse targeted jobs and post your CV for potential employers to view. Visit www.aerosocietyjobs.com for more information.

Become a registered engineer

The RAeS is a licensed body of the Engineering Council, accredited to award the professional qualifications of Chartered Engineer (CEng), Incorporated Engineer (IEng) and Engineering Technician (EngTech).

Becoming professionally registered at EngTech, IEng or CEng level demonstrates you have achieved high-level skills and competencies and are committed to your career as a professional engineer.

Professional registration will ensure you are internationally recognised and rewarded in a highly competitive environment, so join the Society today to continue developing your engineering career.

Insight into technical issues

AEROSPACE is the flagship monthly publication of the Royal Aeronautical Society and is free to all members as part of their membership. It is written for aerospace professionals by aerospace professionals.

AEROSPACE is a features-led magazine, global in outlook, dynamic and forward looking. Each issue contains news and analysis as well as RAeS news and events, book reviews and news of members. As part of your free student membership of the society, you will receive electronic access to the magazine, to download and read wherever you are.

A global network of aerospace professionals

Lara Small ARAeS, Aerospace Engineer, Rolls-Royce says 'I love my membership of the RAeS because of the phenomenal number of opportunities I can take to expand on what I know of the aerospace industry, through informative lectures, inspiring conferences, influential events and modern publications.

'Being a member of the RAeS means that I have access to a powerful community of like-minded professionals at various stages in their careers where I can seek advice for my benefit or contribute to the development of others.'

For more information on how to join, contact our membership team on:

Tel: +44 (0)20 7670 4355
Email: membership@aerosociety.com
www.aerosociety.com

Case study

The AeroMSc Bursary Scheme: launch your career in UK aerospace

If you will graduate in 2014 with a good STEM degree, why not consider aerospace engineering? 'A degree in maths, physics or a related science, technology or engineering discipline is ideal,' said Angela Ringguth, RAeS Project Manager of the Aero MSc Bursary Scheme, 'because it gives the theoretical foundation on which to build practical and specialised skills the leading employers are looking for.'

If your application is successful, the scheme will pay tuition fees, up to a maximum of £9,500 for your chosen MSc – the list of eligible programmes is shown on the website. First of all you need to research the sector a bit and decide which of the six main disciplines interests you and will fit best with your knowledge and interests so far. They are structures, systems, manufacturing, maintenance, aerodynamics and propulsion. Then apply to your chosen MSc programme.

You can submit your application to the bursary scheme while you are waiting for your offer of an MSc place. Competition has been fierce in the first round, so you are advised to get your application in as soon as possible. Experience shows that applicants who have already investigated the sector stand a much better chance of being successful, so you are recommended to contact one of the bursary scheme partners to obtain a letter of support from them – their contact details are given in the FAQ section of our website (**www.raeng. org.uk/aeromsc**).

Background

UK aerospace is a thriving place to make your engineering career. It has a 17% share in the global aerospace market, making it the largest in Europe and second only to the US worldwide. The sector contributes to more than 10% of the GDP attributable to manufacturing, so ensuring an excellent supply of highly skilled and competent aerospace engineers is key for its ongoing success.

Through the Aerospace Growth Partnership, leading aerospace companies have formed a partnership with government to offer 500 ambitious employees and new graduates financial assistance to study master's (MSc) level degrees in aerospace engineering. The competition for these bursaries opened in February 2013 and will run for three years. It is being run in collaboration with the Royal Academy of Engineering and the Royal Aeronautical Society to help the sector develop the high-level skills it needs to compete globally. UK companies sponsoring the scheme are BAE Systems, Bombardier Aerospace Belfast, EADS/Airbus,

Finmeccanica UK, GKN, MBDA Missile Systems, Messier-Bugatti Dowty, Rolls-Royce and Spirit AeroSystems.

The first 100 bursaries have already been allocated, and 19% of these have gone to women. This is excellent news as the engineering professions aspire to attract more women into these careers and to see them progress to the highest levels. Today only 6% of the UK's engineering workforce is female compared with 25% in Sweden and up to 30% in some eastern EU countries.

Main sectors

The industry is dominated by civilian and military sales in approximately equal measure. The space sector accounts for only about 4% of sales.[61]

Civilian

This sector is dominated by aircraft engines (c. 30%), systems and frames (c. 40%) and equipment (c. 25%).[60, 61] There is also a service industry for the maintenance of commercial and private aircraft.

Examples of products include:

- aircraft wings (www.eads.com)

- engines (www.rolls-royce.com)

- fuselage and wings (www.belfast.aero.bombardier.com)

- helicopter transmission and blades (www.agustawestland.com)

- light aircraft (www.britten-norman.com/islander).

Military

Military aircraft sales have fluctuated about the 50% of sales mark over the past decade but are set to decline sharply and for the long term as a result of cut-backs in military spending in the USA, UK and elsewhere.[62] Missiles constitute about

Join the world's foremost aerospace community

ROYAL AERONAUTICAL SOCIETY

The Royal Aeronautical Society is the world's only professional body dedicated to the entire aerospace community.

Established in 1866 to further the art, science and engineering of aeronautics, the Society has been at the forefront of developments in aerospace ever since.

What we do

- **Promote the highest possible standards in all aerospace disciplines**
- **Provide specialist information and act as a forum for the exchange of ideas**
- **Play a leading role in influencing opinion on aerospace matters**

Why become a member?

- Membership grades for every stage of your career
- Professional recognition and development
- Free careers guidance and advice
- The opportunity to contribute to advancing the aerospace profession
- Interact with a global network of approximately 20,000 members in over 100 countries
- Access a wealth of information through monthly publications, the website, social media and the internationally renowned National Aerospace Library
- Preferential rates for the Society's 450+ events and conferences held each year for your continuing professional development

www.aerosociety.com/membership

membership@aerosociety.com

4% of UK aerospace sales. Many companies produce aircraft for both civilian and military use and therefore any engineer wishing to avoid employment on military applications will find their choices more limited.

Examples of products include:

- Apache AH-Mk1 helicopter (www.agustawestland.com)

- Eurofighter Typhoon combat aircraft (www.baesystems.com)

- warfare training (www.cobham.com).

Space

Much of our modern way of life is dependent upon space satellite technology. We use it for communications, navigation and monitoring the planet. Applications include information gathering for security, environmental monitoring, and humanitarian and disaster relief. The UK is the leading exporter of satellite-based tools to the developing world.[63] Our dependency on space technology and the services that it enables will continue to grow and the UK government has committed significant funds to its development in partnership with the European Space Agency. The UK is noted for its satellite telecommunications and by the end of 2013 every UK satellite TV channel will be delivered via a UK-built space craft. Although a relatively small part of the industry, the UK space sector contributes £9.1 billion to the economy and plans to grow to become a £30 billion industry by 2030.[64]

Skills and values required

In addition to the skills discussed in Chapter 4 the aerospace engineer will need to have a passion and good knowledge of the aerospace industry and pay meticulous attention to detail in an industry that is safety critical and operating to very fine tolerances.

Many employers will expect a willingness to work on military applications and where this is the case candidates will need to pass security screening. Candidates may also find foreign language skills useful in multinational companies or where work overseas may be required.

Entry routes

A university degree or equivalent is held by 39% of employees in the UK aerospace industry.[61] The normal route into this field is through studying a degree in aerospace or aeronautical engineering, although employers will recruit from other relevant degrees such as:

- computer science

- electrical and electronic engineering

- general engineering

- mathematics

- mechanical engineering

- physics

- manufacturing engineering

- space-related courses

- software engineering

- systems engineering.

The Association of Aerospace Universities website provides a list of aerospace degree courses within the UK. In 2010/11 approximately 2,500 students commenced courses in aerospace in the UK and between 2004 and 2011 aerospace courses awarded degrees to approximately 1,000 undergraduates per year.[3]

Graduate entry roles

Smaller companies, such as a specialist supplier, will typically provide training on the job supervised by an experienced engineer. Large aerospace companies, such as BAE Systems (www.baesystems.com), will run structured graduate training schemes designed to provide graduates with experience in different areas of aerospace engineering, including their business functions. Multinational companies or groups, such as EADS (www.eads.com), may include a placement overseas. On completion of the training, which may take two or three years, the graduate will usually specialise in an area of the business, such as:

- manufacturing: transforming designs into high-performance manufactured components to meet the required precision, quality and cost constraints

- structural engineering: analysing the designs of structural components and assemblies for stress distribution, dynamic behaviour, fatigue and fracture

- design: from concept to delivery of a prototype design, the engineer will use their knowledge of the design process to deliver solutions to the brief they have been given using sophisticated computer visualisation tools and software

- aerodynamics: studying the flow around the aircraft and modelling it with computational fluid dynamic software with the aim of achieving a better performance and reduction in fuel consumption and noise

- system simulation: from control systems through to complete flight simulators used to train pilots.

Who will I work with?

Graduates working in the aerospace industry will work with technicians on the shop floor, project managers, research engineers, customers, supply chain companies providing components or raw materials, licensing authorities, operators and staff working in the business support functions.

Salary levels

The average starting salary in 2011 was £25,961 for aerospace graduates and is slightly higher than the average for all engineering and technology graduates of £25,762. The average aerospace salary in 2011 was £37,391.[3]

Graduate starting roles in 2013 are advertised in the range £20,000–£28,000 depending on company and location. A number of larger companies also make a welcome payment of typically £2,000–£3,000 (e.g. BAE Systems and Airbus). For engineers with experience, salary ranges of £25,000–£45,000 are advertised. Salaries for senior levels will exceed £50,000.

10

Materials engineering

'Materials Science and Engineering involves the study of the structure, properties and behaviour of all materials, the development of processes to manufacture useful products from them, and research into recycling and environmentally friendly disposal.'[65]

'Materials Scientists or Engineers look at all of the different groups of materials, metals and alloys, polymers, ceramics and composites. They develop new materials for new applications, improve existing materials to give improved performance and look at ways in which different materials can be used together.'[66]

In any engineering endeavour success is linked to the nature of the materials selected and used because of their properties and behaviour and it is these that give the material value. Examples of material types are:

- ceramics: used for example in crucibles for molten metal or as electrical insulators

- chemicals: a wide range of industrial chemicals are manufactured and sold

- composites: different materials combined together to benefit from the properties of each constituent

- glass

- metals

- minerals: industrial minerals are natural products used in a wide range of applications. For example bentonite clay is used to provide temporary support to excavations in the ground (e.g. trenches, tunnels, boreholes); manufacture of cat litter; and for facial mud packs

- plastics

- polymers: both natural (e.g. rubber) and synthetic (e.g. nylon, PVC and polypropylene)

- textiles: industrial applications include reinforcement (e.g. pipes and geotextiles) and fire protection.

There are a number of properties for which a material may be selected:

- strength: how much stress can the material sustain without failing?

- stiffness: how does the material deform under load?

- electrical properties: is the material a good conductor or insulator?

- resistance to heat: can the material withstand the operating temperature, for example in a gas turbine?

- durability: does the material degrade in the environment in which it is used?

- weight: are lightweight components required, for example as a means to reduce fuel consumption?

Consideration must also be given to cost, both short- and long-term. The sustainability of material resources and their use in a responsible manner for the well-being of society are increasingly important considerations.

The properties of materials are linked to the microstructure of the material, including which atoms are present, how they are joined, and how groups of atoms are arranged throughout the material. Material engineers can manipulate the resulting properties through the way in which the material is made and processed.

The UK Centre for Materials Education attributes the last 20 years of world-transforming technological advances to developments made in materials science and engineering and expect this to continue: 'Materials Science and Engineering has become a key discipline in the competitive global economy and is recognised as one of the technical disciplines with the most exciting career opportunities.'[65]

The nature of the industry

Material engineers work to provide better material solutions and performance. They may be engaged in developing or harnessing new materials to meet their requirements or providing improvements in the use, processing and cost-effectiveness of existing materials. The UK Centre for Materials Education states that: 'Materials are evolving faster today than at any time in history; enabling engineers to improve the performance of existing products and to develop innovative technologies that will enhance every aspect of our lives'.[66]

As such, material engineering is an enabling discipline that supports a wide variety of activities. Materials UK identified four themes in their 2008 review.[67]

1. Tackling climate change through improvements to the efficiency of energy generation, use of renewable technologies, energy conservation and sustainability of production and consumption methods.

2. The need for materials to meet the demands in the construction industry, both domestic and commercial. As well as the traditional construction materials (e.g. brick, timber, concrete and steel), the industry is investigating new construction materials such as the

lightweight pultruded fibre reinforced polymer sections recently used to construct a test house in the Startlink Composite House Project.[68] Other materials include those for services, insulation, cladding, etc.

3. Major industrial sectors such as automotive, aerospace, oil and gas, chemical and marine.

4. Other activities in which the UK is strong, including the pharmaceutical and biotech, food, ICT/multimedia, fashion, and design industries.

Main sectors

Materials engineers are employed across many manufacturing and industrial sectors. The sectors described here are those identified in the report jointly prepared by the European Materials Research Society and European Science Foundation entitled *Materials for Key Enabling Technologies*.[69]

Advanced materials

The development of advanced new materials and techniques has important application in industries where innovation and high value-added engineering provides the competitive edge. The UK government believes that developments in new materials over the next 20 years may match the benefits gained from the increased variety of plastics developed in the twentieth century.[70] Here are some examples.

- **Aerospace:** where the targets are reduced weight and increased operating temperatures in the turbines to improve fuel consumption. For example ceramic matrix composites (CMCs), where the ceramic matrix is reinforced by ceramic fibres, achieve improved thermal and mechanical properties. This makes them suitable for use in the hot section of aero engines with a potential increase in turbine inlet temperature of 25%, achieving a potential 6–8% increase in fuel efficiency. A useful discussion on this and other advanced materials in aerospace is provided in a review authored by Sir David King *et al.*[71] Aerospace engineering is discussed more fully in Chapter 9.

- **Automotive:** using new materials to improve fuel efficiency, reduce emissions and improve safety and robustness of the vehicles. One example is the development of so-called 'self-healing' materials that when damaged trigger a repair process. In simple terms this is achieved by encapsulating chemicals within the material such that when damage occurs they are released and are free to mix and react to create the repair. Another example is a coating for metal panels that when scratched self-heals to prevent corrosion to the panel.[72]

- **Biomedical:** encompasses advanced materials for implants and the scaffolds that service the development of cells and tissue. These tissue scaffolds must allow cells to migrate through and colonise the structure. They must also allow delivery of nutrients and removal of waste from the cells. Future developments may well lead to on-demand printing of customised implants in hospitals, using biological materials. The combination of intelligent implants to monitor and diagnose a patient's condition, improved delivery of drugs and artificial replacement organs promise increased life expectancy.[73]

- **Textiles:** examples of current developments include the following:[74]

  ○ Anti-microbial treatments that target undesirable organisms whilst remaining safe, non-toxic and compatible with the other textile treatments. The application of the antibacterial agent should not impact negatively on the textile properties and appearance or the environment.
  ○ Treatments for flame-resistant or retardant textiles used for clothing, furnishings and applications such as parachutes. Flame-retardant textiles reduce the ease of ignition and propagation of the flame, whilst flame-resistant textiles also form a barrier to flame penetration, for example to the person wearing the garment.

Materials for energy

With an emphasis on harnessing renewable energy sources such as wind, tidal and solar there is a need for materials capable of efficiently converting the natural energy into electric energy. For example wind turbines must be aerodynamically efficient and be manufactured from materials of adequate fracture strength and toughness. The electrical energy generated may be stored as electrochemical energy, requiring an understanding of catalytic reactions and thus surface and interface phenomena within the cell. Improved design and materials are also needed to improve the sustainability of rechargeable batteries used to power mobile technologies such as phones, tablets and computers.

Nanotechnology

The 2010 government report, *Technology and Innovation Futures: UK Growth Opportunities for the 2020s*, identified a $100 billion market for nanomaterials.[70] Nanomaterials are those materials composed of particles with one or more external dimensions in the size range from 1 to 100 nanometres (please refer to the Glossary for the full definition). There are some 200 companies in this sector, with strengths in the development of coatings, composite materials and nanomaterials, medical technologies, and displays and sensors.[75] An example of an application is Solaveil, a thin layered material manufactured by adding carbon nanotubes to thermoplastic fibres. The material is printed and applied to window glass to control solar radiation and glare. A variation on the technology is a surface application for use on hospital windows that resists *Staphylococcus aureus* and *E. coli*.[76]

Materials for micro-, nanoelectronics and silicon photonics

Photonic elements, which convert electric signals into light and back into electric signals, are used for communication systems. In the future they will increase the speed of electronic systems as they are used for data transmission within circuit boards and processors. Electric lighting will increasingly use light emitting diodes.[69]

Quantum dots are nanoparticles that emit photons under excitation. These are visible to the human eye as light. By controlling the size of the quantum dots it is possible to control the wavelength of the photon emissions and hence the colour. The smaller the dot, the closer it is to the blue end of the spectrum, and the larger the dot, the closer to the red end. Dots can be sized to emit photons in both the infra-red or ultra-violet ranges. Quantum dot technology can be used for everything from lasers to domestic lighting. Quantum dot lasers are cheaper and offer a higher beam quality than conventional lasers, whilst quantum dot LEDs have a lifespan of up to 15 times that of energy-efficient bulbs with a similar energy efficiency. The technology is also used for street lighting, signs, etc. where energy efficiency and longevity are desired.[77]

Biotechnology

The biotechnology sector has exploited the properties of nanomaterials for applications such as:

- analytical tools: using a nanopore or nano-scale hole to isolate an individual molecule for identification[78]

- contrast agents: used to enhance the contrast when imaging cell structures

- diagnostic devices: for example lab-on-a-chip technology for faster, at point of use diagnostics

- drug delivery vehicles: transporting the drug to specific cells and in the process reducing drug consumption, side-effects and cost.[79]

Entry routes

The normal entry route is via an accredited degree in materials science or engineering. Materials engineering roles exist in a range of industries and therefore graduates from other disciplines with an understanding of materials can enter these roles, such as applied sciences, metallurgy and other engineering degrees.

Graduate entry roles

The many opportunities for graduates reflect a skills shortage in this area and roles will vary according to the sector, product and size of company. Materials engineers may choose to work in research and development or focus on the production and processing side. The types of roles that a graduate might tackle include:

- feasibility studies for adapting existing or installing new equipment and/or processes

- material selection to meet a project specification

- materials testing to establish properties

- modelling the life cycle of a product and material and identifying how materials might be recycled at the end of the product's useful life

- monitoring compliance with safety legislation and best practice

- troubleshooting when processes, materials or products do not perform as intended

- working on prototype designs.

Salary levels

Graduate positions in 2013 were advertised with starting salaries in the range £20,000–£28,000 depending upon the employer and location. The website prospects.ac.uk[80] reports salary ranges of £27,000–£40,000 with some experience and an ability to earn up to £60,000 with 10–15 years' experience.

My story . . .

Mukunth Kovaichelvan, Materials and Process Modeller at Rolls-Royce

I graduated with the degree of MEng Mechanical Engineering in 2009 and joined Rolls-Royce as a graduate engineer. During my degree I had undertaken a one-year industrial placement in operations management at a GE Energy's industrial gas turbine repair and overhaul facility, which helped me secure my graduate position. I completed an 18-month manufacturing engineering Professional Excellence Graduate Training Scheme at Rolls-Royce, which consisted of six three-month long placements across the company in the following departments:

- Precision Casting Facility

- Strategic Research Centre

- Future Programmes Engineering

- Manufacturing Technology

- Design and Make

- Combustion and Casings Thermals.

During the placements I was given a number of projects to work on, such as:

- implementing a statistical process control to manage the raw materials in the foundry

- conducting a feasibility study of novel offshore wind turbine technologies using specialist software

- performance modelling and optimisation of a new large aero-engine

- establishing a standardised process within the Rolls-Royce production system for testing tooling and fixtures across all of the manufacturing facilities in the company

- the design, manufacture, testing and analysis of a flow visualisation rig

- analysing the thermo-mechanical behaviour of a turbine casing.

In my current role as a Materials and Process Modeller I lead projects aimed at developing computational methods to model manufacturing processes, working closely with factories, universities and research centres across the country. My role sits in the middle of modelling material behaviour, the physics involved in manufacturing using those materials and the complexities of designing components that exploit these properties. The sheer variety and complexity of the technical challenges at hand could keep my brain engaged for a lifetime.

11

Environmental engineering

'Environmental engineering is concerned with the measurement, modelling, control and simulation of all types of environment. It is an interdisciplinary subject, bringing together aspects of mechanical, electrical, electronic, aeronautical, civil, energy and chemical engineering. It also draws from the fields of physics, acoustics, metallurgy, microbiology, pharmacy and many other technical and scientific disciplines.'[81]

'Environmental engineering applies the theoretical background provided by the environmental sciences using engineering principles. The aim of an environmental engineer is to sustain or improve the natural environment in order to guarantee a good and healthy life quality for human beings as well as for other organisms.'[82]

Environmental engineering as a title may be considered ambiguous by some, particularly when the nature of the work undertaken proves to be so varied and with different purposes. The definitions above suggest that the discipline is primarily focused on the engineering of the natural environment to secure its future and the well-being of those who occupy it. In this sense the discipline encapsulates the principles of sustainability and the engineer's focus on minimising waste and limiting the adverse impacts of their work on the environment, but equally it may suggest a proactive intervention to manage the environment, as for example in prevention or control of extreme flood events. With a burgeoning world population, the great global challenges of protecting the natural environment, providing adequate drinking water and power all sit at the heart of environmental engineering. Many environmental engineering degree courses focus on the natural environment and will include modules on the legal and technical aspects of control and remediation of pollutants in the ground, air and water. There is a natural synergy between environmental and civil engineering that is often reflected in these courses.

This focus on the natural environment overlooks the activity of environmental engineers who seek to control the built environment for the benefit of those who live and work there. Their activity will include the design of services within buildings to both achieve the desired functionality and to promote the health and happiness of those who occupy them; and may also extend to the design of the urban landscape to assist groups with particular needs. These skill sets will be developed in building services, civil, mechanical and architectural degree courses.

The third area of work captured under the heading of environmental engineering is that of engineers whose focus is the design of equipment to function as expected within the environment within which it will be used, particularly where the equipment may be susceptible to dust contamination, extreme temperature, vibration or electromagnetic interference. This is a specialist skill set that draws upon the knowledge of mechanical, electrical and electronic engineers.

The nature of the industry

All engineers, irrespective of their roles, should have a concern for the environment and the potential impact of their work upon it. This principle is captured in the various codes of ethical conduct promoted by engineering organisations. For example in 2007 the Engineering Council UK and the Royal

Academy of Engineers jointly published a Statement of Ethical Principles that states:

> *'Professional Engineers . . . should:*
>
> * *minimise and justify any adverse effect on society or on the natural environment for their own and succeeding generations*
> * *take due account of the limited availability of natural and human resources*
> * *hold paramount the health and safety of others.'*[83]

An earlier Code of Environmental Ethics for Engineers published by the World Federation of Engineering Organizations (WFEO) in 1986 provides an even more forthright assertion of the engineer's responsibility:[84]

> *'The WFEO Committee on Engineering and Environment, with a strong and clear belief that mankind's enjoyment and permanence on this planet will depend on the care and protection provided to the environment, states the following principles:*
>
> *To all engineers when you develop any professional activity:*
>
> 1. *Try with the best of your ability, courage, enthusiasm and dedication to obtain a superior technical achievement, which will contribute to and promote a healthy and agreeable surrounding for all, in open spaces as well as indoors.*
>
> 2. *Strive to accomplish the beneficial objectives of your work with the lowest possible consumption of raw materials and energy and the lowest production of wastes and any kind of pollution.*
>
> 3. *Discuss in particular the consequence of your proposals and actions, direct or indirect, immediate or long term, upon social equity and the local system of values, and upon the health of people.*

4. Study thoroughly the environment that will be affected, assess the impacts that might arise in the state, dynamics and aesthetics of the ecosystems involved, urbanised or natural, as well as in the pertinent socio-economic systems, and select the best alternative for an environmentally sound and sustainable development.

5. Promote a clear understanding of the actions required to restore and, if possible, to improve the environment that may be disturbed, and include them in your proposals.

6. Reject any kind of commitment that involves unfair damage to human surroundings and nature, and negotiate the best possible social and political solution.

7. Be aware that the principles of ecosystemic interdependence, diversity maintenance, resource recovery and interrelational harmony form the bases of our continued existence and that each of those bases poses a threshold of sustainability that should not be exceeded.

Always remember that war, greed, misery and ignorance, plus natural disaster and human induced pollution and destruction of resources, are the main causes of the progressive impairment of the environment and that you, as an active member of the engineering profession, deeply involved in the promotion of development, must use your talent, knowledge and imagination to assist society in removing those evils and improving the quality of life for all people.'

It is therefore unsurprising that a demand for a cadre of specialists in environmental engineering has developed across many of the other engineering industries described in this book, driven also by increasing legislation designed to protect the environment. That the demand for such skills is set to grow was recognised in the Engineering UK 2013 The State of Engineering Report,[3] which observes, 'As world consumption continues to grow, there is a growing risk of resource shortages (or price hikes in anticipation) in a wide range of commodities. New jobs in recycling and resource conservation will emerge.'

Environmental engineers will be engaged in a number of areas.

- Consultancy: providing environmental solutions to companies, local and national government and public services. Examples might include management of waste recycling, increasing energy efficiency or conducting an environmental impact assessment of schemes such as a new rail or road scheme.

- Contracting: providing specific services to clients, such as cleaning up ground pollution associated with past activities on a site. For example an old manufacturing site may well have contaminated the ground with spilt chemicals and waste that need to be treated before the site is used for new housing.

- Manufacture or construction: from design and optimisation of processes to reduce pollution through to monitoring environmental impact to ensure compliance with statutory requirements.

- Public authorities, government departments, executive agencies and utilities: ensuring that the environment and public health are protected, for example by monitoring air and water quality, noise levels and disposal of waste.

- Research and development: of new technologies and processes to restore a damaged environment or reduce the future impact on the environment, for example harnessing renewable energy sources or developing electric vehicles.

Examples of employers include:

- AMEC's Environmental Division (www.amec-ukenvironment.com): consultancy

- Ecosulis (www.ecosulis.co.uk): contractor/consultant

- Environment Agency (www.environment-agency.gov.uk): executive non-departmental public body

- MBDA (www.mbda-systems.com): manufacturer of guided weapons

- Procter and Gamble (www.pg.com): manufacturer

- Thames Water (www.thameswater.co.uk): utility company.

Main sectors

The diversity of industries within which environmental engineers may work makes compiling a comprehensive list of sectors difficult, but the following selection gives a good overview of the types of work undertaken. Some sectors are discussed elsewhere in this book, including building services (Chapter 5) and water engineering (Chapters 6 and 8).

Environmental impact assessment (EIA)

What will be the effect of a project on the environment? EIAs are designed to answer this question by providing a structured review of all the implications and to minimise their impact by considering the means by which they can be mitigated. The EIA requires the engineer to consult with a range of interested parties. The environmental engineer will hope to be engaged in a new project from the earliest stages of design because this is the period during which modifications to the proposal can be incorporated in the most cost-effective manner. Making changes in the later stages of a project can be difficult and the EIA helps to avoid this. Having identified measures to offset the environmental impact they can be formalised into an environmental management plan to inform the implementation stage of the project and any subsequent operational requirements. An example of an EIA project is the modification of the coal fired Eggborough Power Station to tackle emissions of sulphur dioxide. A system was introduced that uses limestone to capture the sulphur in the flue gas and in the process produce gypsum that can be used in the production of plasterboard. The EIA extended from the sourcing of the limestone through to removal of the gypsum. The assessment gave consideration to the impact on 'traffic and transportation, air quality, noise, water quality, ecology, landscape and visual impacts, contamination, health and safety and cultural heritage'.[85]

Energy

Power generation plants are subject to increasing and more demanding legislation to reduce pollution. Where the power station needs modification to meet the legislation an environmental assessment of the technologies on offer is performed as described above. EIAs are required for other forms of power generation including:

- biomass

- energy-from-waste

- landfill gas

- large solar parks

- wave and tidal technologies

- wind turbines (on-shore/offshore).

Environmental forensic engineering

Environmental forensic engineering seeks to identify the source of a contaminant in air, ground or a waterway. If that contamination was caused illegally there will be a requirement to establish robust evidence that can be used within a legal process for environmental litigation.

Geo-environmental engineering

This sector is primarily focused on the use of brownfield sites – sites that have a history of use and potential contamination of the ground. The ground will undergo environmental assessment as a part of the site or ground investigation commissioned for a project. If a problem is identified the ground contaminants are treated according to the nature of the contamination and the risk it poses. The environmental engineer will require knowledge of the contaminants typically associated with different uses of the land and the chemical, biological or physical techniques available to treat them. Contaminants may also arise from old mine workings or historical dumping of waste both legal and illegal. Some pollutants are mobile and will flow through the ground, extending the affected zone and potentially contaminating water courses, posing a health risk for any humans or animals that consume the water. The environmental engineer will model such flows and propose solutions or remedial measures.

Health and well-being

This sector is broad and encompasses three main areas.

- Design of buildings for comfort, including energy-efficient temperature, humidity, noise and light control and creating a sense of individual well-being.

- Design of the urban environment for the well-being of communities and vulnerable groups such as the elderly, children or those suffering with dementia. I'DGO (Inclusive Design for Getting Outdoors), a research consortium specialising in this area, promotes best practice in the planning and design of the urban environment for all users, claiming that, 'There is growing evidence that well-designed outdoor spaces can enhance the long-term health and well-being of those who use them regularly.'[86]

- Pollutant monitoring and control, including water management, noise and atmospheric pollutants from, for example, industrial plants, vehicles and incinerator plants.

This sector provides for an assessment and management of those factors affecting human health by the environmental engineer through an awareness of:

- architectural design for well-being

- engineering risk management

- environmental epidemiology

- environmental toxicology

- pollutants and their control.

Reliability of products and structures

Environmental Engineering, the journal of The Society of Environmental Engineers, covers topics that are collectively grouped here as reliability of products and structures. This captures the work of environmental engineers that is not focused on the natural or urban environment but is concerned with the following.

- **Climatics:** ensuring that the equipment will function reliably in the environments that it will be exposed to, including temperature, pressure, moisture, corrosive environments, sand, dust, ice and solar radiation.

- **Condition monitoring:** instrumenting the equipment or structure to monitor how it is performing and to trigger actions such as maintenance before failure occurs. Examples include monitoring engines or vehicles and monitoring structures such as bridges with embedded and surface mounted sensors.[87] Researchers are developing new techniques such as smart paint that can detect microscopic faults in wind turbine foundations and bridges. The paint is made from fly ash and highly aligned carbon nanotubes and can be interfaced with a wireless sensor network for remote monitoring.[88]

- **Contamination control:** of surfaces and atmosphere in applications sensitive to dust, microbes or environmental variation. Industrial applications in, for example, the electronics, pharmaceutical and biotechnical industries will require use of clean rooms and/or bacteria-resistant surfaces.

- **Electromagnetic compatibility:** electromagnetic emissions may cause electromagnetic interference that can disrupt the performance of other equipment, therefore engineers must control the emissions and also test the resilience of equipment operating in the presence of electromagnetic disturbances. Electromagnetic emissions from data signals transferred via cables and wires may constitute a data security risk if intercepted, and engineering solutions to protect sensitive data will be employed by organisations such as the military and banks.

- **Vibration, shock and noise:** testing equipment to ensure that it can withstand both regular and unusual exposure, for example how does a computer or smartphone cope with being dropped?

Security

What is intended by environmental security? Michael Renner notes that there are two approaches to understanding what is meant by 'security':

> *'Today, there are essentially two major conceptions of human security. The first approach focuses primarily on protecting people from acts of violence and violent threats to their rights, safety, or lives – "freedom from fear". . . . The second approach stresses far broader issues of human well-being and dignity and might be characterised as "freedom from want". It focuses on protecting people not only from violence but also from a far more expansive array of social, economic, and environmental challenges.'*[89]

In terms of the latter approach, Renner identifies the following key issues.

- Energy: in particular oil, gas and coal resources. These are characterised by inequality of consumption, militarisation of oil-rich nations and the impact of carbon emissions for the future.

- Food security: is threatened by water shortage, environmental degradation and a loss of biodiversity. Growth in global population is predicted to increase food demands by 70% by 2050.[90]

- Infectious diseases: spread quickly and impact on more people as a result of growing migration, international travel and refugee displacements. Climate change and engineering works such as the building of reservoirs can exacerbate the problem.

- Water: irrigation and food production uses approximately 70% of global freshwater withdrawals. Increasing demand for food and diversion of water to growing biofuels will add to water shortages.[90]

All of these issues have the capacity to trigger conflict with its repercussions felt across the globe. This is a challenge that is as much a political problem as it is an engineering problem, but it does drive a concern for protecting the environment and living more sustainably in order to secure our future. Environmental engineers have a readily apparent role to play in responding to these challenges.

Waste

With the emphasis being increasingly placed on sustainability and the government using taxation to discourage disposal to landfill, waste is increasingly recognised as a resource. Environmental engineers will:

- seek to design waste out of a project as far as is practicable

- identify uses for the waste to avoid sending it to landfill

- design and operate recycling and recovery schemes

- design and operate waste tips

- ensure that industrial waste is collected, stored and disposed of in accordance with the legislation for the type of waste being handled.

Entry routes

An accredited degree in environmental engineering or from a relevant discipline such as civil, chemical, materials, manufacturing or mechanical engineering are acceptable routes into this industry.

Graduate entry roles

As a graduate engineer your work will depend upon the type of company and industry that you enter. Typical activities include:

- assessing sites for contamination, e.g. brownfield or industrial

- conducting technical audits

- designing and implementing remediation or reclamation schemes

- designing, developing, testing and implementing technical solutions to environmental problems

- environmental impact assessments

- environmental testing of equipment and products

- monitoring compliance with environmental and safety regulations

- monitoring the performance of equipment and structures

- resource management

- safety and risk management

- waste management schemes.

Graduates with appropriate work experience may seek to become a chartered engineer through one of the Engineering Council UK's professional engineering bodies or a chartered environmentalist through the Society for the Environment's licensed member organisations (www.socenv.org.uk/members) as described in Chapter 16.

Salary levels

Graduate positions in 2013 are being advertised with starting salaries in the range £18,000–£25,000 depending upon the employer and location. The website prospects.ac.uk[91] reports salary ranges of £20,000–£30,000 with some experience and an ability to earn up to £50,000 with experience of 10 or more years.

My story . . .

Leon Hawley, Assistant Engineer at URS Corporation

Upon completion of my Postgraduate Diploma in Environmental Management in 2005 I joined a small environmental consultancy. This consultancy specialised in remediation of contaminated sites. I worked largely in the field of contaminated land, undertaking Phase 1 Desk Studies and undertook ground investigations involving site supervision and sampling and analysis of chemical testing of environmental samples.

In 2006 I joined GIP Ltd, which was a ground investigation contractor. Here I gained useful field and project management experience. I designed and managed site investigations including:

- logging of soil and rock samples

- producing reports of logs that complied with the British standards

- locating boreholes and trial pits for in-situ monitoring of ground conditions

- designing and supervising contaminated land investigations, including scheduling laboratory geotechnical and chemical testing and in-situ monitoring

- undertaking qualitative geo-environmental risk assessments.

It was at this stage I wanted to add more design experience to my career, so in 2007 I joined the engineering consultants Waterman as a graduate engineer within their geo-environmental team. Here I processed and interpreted site data and prepared reports, making design recommendations for:

- earthworks

- foundations

- drainage solutions

- mitigation measures for contaminated land and landfill gas for residential and industrial development sites

- environmental risk assessment, including human health and controlled waters risk assessment, ground gas assessment, hazardous waste classification.

I worked on large development projects and was resident engineer on Bilston Urban Village, a large remediation and earthworks project, for four months. I also planned, managed and supervised ground improvement and earthworks for development sites to create developable plateaus.

In 2009 I joined URS where I began working on dams and reservoirs as well as geotechnical projects. I have worked under senior reservoir engineers as mentors to inspect dams and reservoirs around the UK as part of the statutory requirements for regular safety inspection. I was also involved in a £5.5 million local project at Chasewater Reservoir, which involved earthworks design, spillway and overflow design and project management. I have learnt a great deal about geotechnical and hydraulic design during my work with reservoirs and am about to undertake the design of a 50,000 cubic metre lake in 2013. I have also been overseas to Azerbaijan and Georgia to undertake ground investigation for the proposed BP gas pipeline across the two countries. I currently work on a wide range of projects from ground investigations, earthworks contracts, slope stability analyses, retaining wall design and dam design and project management of remedial works to dam structures.

I initially found that a lot of the material covered within my environmentally based degrees was not relevant within the scope of work of an engineering consultancy. Working with contaminated land, I had found elements of my degree useful, but there was a lot of wider engineering and geological knowledge that was needed and that I did not have at first. The practical on-site work was also something I had never previously experienced. It was a steep learning curve and I spent a lot of time after work trying to get to grips with what I had done during the day. The technical engineering equations were a particular challenge as I had not studied mathematics or physics at A level. Fortunately I had friends and family who were able to explain the initial geotechnical calculations I needed.

When I moved into a design consultancy in 2007, I found myself in a geo-environmental role which combined elements of my environmental background with geotechnical engineering and geology. It was at this time that the work we were doing involved detailed geotechnical design as well as contaminated land remediation. I found that I would only be put forward for the environmental part of the project and the geotechnical part of the project was then given to another employee with an engineering degree. It is common that environmental consultants do initial pre-planning and feasibility assessments but have no involvement once the project work is underway. I wanted to be involved with the whole project and therefore decided to increase my knowledge base. I was able to attend relevant modules from an accredited degree course in civil engineering at a local university to gain an understanding of engineering principles that I needed.

There is a lot of support for training in employment and as a part of The Institution of Civil Engineers' training agreement you continue to gain knowledge. I have attended numerous training courses in support of my work and as part of continuous professional development I have attended evening talks given by experienced engineers and undertaken wider reading and research into developments within the industry.

When you get to a stage in your engineering career where you are in a design office and your drawings are being sent to contractors on site for construction, it is an exciting moment. It is unlikely the design is all your own work because it has gone through a series of checks and changed numerous times, but despite this when built it is something you had a hand in creating. Working for a multinational corporation gives you access to large, high-profile projects such as Crossrail, HS2 and the Birmingham Airport runway, all of which have passed through my office in the last year.

In my work nothing is repetitive; I get opportunities to work on different types of project and to travel around the UK and overseas. I try to spend 35% of my time out of the office to 'mix things up' and give me experience on different sites. For example this year I have been working on the Mersey estuary collecting dredging samples for a new container port in Liverpool. This diversity in jobs and locations is very exciting at this stage of my career as I gain as much experience as possible before applying to become a chartered engineer.

12

Manufacturing engineering

'We define manufacturing engineering as the engineering approach to all factors involved in the production of goods or services. To become broader in their knowledge of all the elements of the manufacturing process, manufacturing engineers must become more involved in understanding customers' needs and desires, and influencing product design to ensure the production of high-quality, low-cost, finished items.'[92]

Manufacturing engineers work in a wide range of industries to make the goods that underpin our lifestyle. Their products include the goods we buy and use, but also the machines and systems that were used to manufacture or process them before they became available to us. Manufacturing engineers focus on the systems and processes of manufacture in order to efficiently produce the required goods to the quality required. They are concerned both with the machinery or processes that are used and also the arrangement of these elements to create manufacturing systems. For example in a car plant the manufacturing engineers will design the production line taking raw materials or component parts and producing the finished product, giving consideration to quality control, safety of the production line workforce, manufacture or assembly techniques, material handling and stock control. These skills for optimising processes are now finding use elsewhere in, for example, the design of hospital systems for patient care.

The nature of the industry

Manufacturing often has a poor image created by press stories about company closures and statistics about the transfer of manufacturing to other countries. Certainly manufacturing is sensitive to global and local economic pressures and has halved in size between 1990 and 2010 according to the Office for National Statistics. However, the reality is that nearly 3 million people work in manufacturing industry and manufacturing provides 10% of the UK's gross value added (GVA), a measure of the value of goods and services produced. In 2012 the UK was ranked ninth in the world for the value of manufactured output.[93] The government have prioritised support for the resurgence of manufacturing and in particular high-value manufacturing.[94] This distinction is important because it differentiates the areas of manufacturing in which the UK maintains an advantage over countries offering low-cost manufacturing. But what constitutes high-value manufacturing? Sir John Rose, Chief Executive of Rolls-Royce, identifies the following characteristics of high-value activity:[5]

> 'It involves what I call deep knowledge:
>
> - It has a high research and technology content;
>
> - It requires a profound understanding of the customer;
>
> - It exploits both scientific and experiential intellectual property;
>
> - It involves the definition of solutions that meet complex requirements;
>
> - It requires well-developed systems integration skills;
>
> - It involves managing data to inform responses to complex events; and
>
> - It is difficult to do well.'

Examples of employers of manufacturing engineers include:

- BAE Systems (www.baesystems.com): aerospace and maritime

- GlaxoSmithKline (www.gsk.com): pharmaceutical and healthcare

- Jaguar Land Rover (www.jaguarlandrover.com): automotive

- JCB (www.jcb.co.uk): specialist automotive

- Kraft Foods (www.kraftfoodscompany.com): food and drink

- Morrisons (www.morrisons.co.uk): food processing and retail

- Rolls-Royce (www.rolls-royce.com): aero-engines, marine and power generation

- Toyota (www.toyota.com): automotive.

Main sectors

Manufacturing engineers have roles in most engineering sectors, including aerospace, chemical, materials and mechanical engineering. The latter are discussed in Chapters 9, 8, 10 and 5 respectively. Manufacturing embraces a wide range of industrial applications beyond the commonly recognised industries; examples include pharmaceuticals, processed foodstuffs and electronic gadgets. Table 4 shows the gross value added (GVA) and numbers employed in manufacturing sectors.

TABLE 4: Manufacturing gross value added (GVA) and employment

Sector	GVA (£ millions)	Numbers employed (000s)
Metal, plastic and non-metal mineral products	28,005	584
Food, beverages & tobacco	27,771	399
Machinery, electrical and transport equipment	22,748	412
Other manufacturing	21,046	566
Chemicals	16,926	566
Pharmaceuticals	10,023	38
ICT and precision instruments	8,393	138
Automotive	6,955	133
Aerospace	5,610	112
Shipbuilding	1,246	32

Source: BIS Economics Paper No. 18: Industrial Strategy UK Sector Analysis[39]

The following are examples of the problems tackled by manufacturing engineers.

- Computer aided systems for translating designs into manufactured articles through the control of equipment and processes. Designs are developed as digital models using software and used to generate instructions to robots and machine processes such as drilling, cutting, turning, welding, etc. to manufacture the finished component.

- Design for manufacture, ensuring that the design of the component is influenced by the company's ability to manufacture it to appropriate tolerances at minimum cost, for example the initial design may have attributes that are expensive or difficult to manufacture and an adjustment to the design may be needed to facilitate the ease of manufacture.

- Logistics, managing warehousing, supply chains and deliveries to ensure that raw materials and products arrive where they are needed on time.

- Performance measurement and improvement techniques such as:

  o balanced scorecard (BSC) used to monitor staff activity against financial and other targets

- ○ overall equipment effectiveness (OEE) based on the components: availability of equipment, performance of workstation compared with the target, and quality or the proportion of units started that are completed satisfactorily.

- Production control techniques such as:

  - ○ enterprise resource planning (ERP) software systems that manage all of the company's resources, information, and business functions from shared databases

  - ○ just in time (JIT) systems that reduce the cost of storing excessive inventory by monitoring their levels and triggering an order for, or manufacture of, new parts when needed

  - ○ materials requirements planning (MRP), normally a software system for managing just sufficient stock levels to ensure that materials and components are available when required for the manufacturing process and that goods are available for delivery to the customer when required.

- Manufacturing resource planning (MRP II) covering all the resources of the company, by extending MRP considerations to include resources such as finance and personnel. The system is designed to facilitate effective and comprehensive decision-making in manufacturing.

- Optimised production technology (OPT) for managing bottlenecks in the production system.

- Quality management techniques including statistical methods and metrology.

- Selection and use of:

  - ○ casting, requiring careful design of moulds and pour sequences

  - ○ joining and cutting processes and equipment

- ○ machining centres

- ○ metal forming equipment, for example presses

- ○ process and light assembly equipment

- ○ robots to control the movement and placing of components and finished articles.

- Waste reduction in the manufacturing cycle.

Skills and values required

The following skills and values for manufacturing engineers extend those discussed in Chapter 4:

- adaptable, flexible and able to work under pressure – for example if the production is halted by equipment malfunction it must be repaired quickly and safely

- using software and other techniques to model manufacturing system layout, performance, output, etc.

- able to design processes and systems with due consideration to the specification, sustainability and safety. Capable of assessing designs for ease, accuracy and cost-effectiveness of manufacture

- able to take a holistic view of the process or system and the various interactions of the factors affecting it.

Entry routes

Accredited degrees in manufacturing or industrial engineering provide the academic component for becoming a chartered engineer, however industry will recruit from alternative degree courses into manufacturing roles, including chemical, electrical, electronic, mechanical and systems engineering where graduates offer the appropriate skillset for their business.

Graduate entry roles

Entry roles in larger companies will be shaped by the company's graduate training scheme which will typically rotate the employee through different functions and departments within the company. The company Rolls-Royce, for instance, runs a Leadership Programme consisting of two stages. In the first stage, graduates are rotated through three six-month placements, one of which will be overseas. The second stage consists of two longer placements of 18–24 months in which the graduate engineer is given greater responsibility and autonomy for a project dependent upon the section of the company in which they are placed. In contrast, Toyota runs an 18-month graduate development programme that starts with a three-month induction, including an eight-week period working in production. The graduate moves from the production tasks into their chosen area of specialisation under supervision and has an opportunity to be seconded to a different area to broaden their appreciation of the company's activities. Upon successful completion they progress to a specialist or engineering role within the company.

Who will I work with?

Given the range of industries that recruit manufacturing engineers a graduate may find themselves as part of a large manufacturing team or in a multidisciplinary team working with other engineers such as chemical, civil, electrical, material, mechanical and systems engineers. They will also interact with the commercial, technical and business functions according to the nature of their employer's business. Given the nature of manufacturing, graduates will have contact with workers on the production line.

Salary levels

According to Prospects starting salaries are typically £22,000–£28,000 depending on size of the company and location. The average starting salary was £26,000 in 2011.[3] Chartered engineers will earn £40,000–£50,000 and experienced senior engineers will earn up to £60,000. Executives can earn significantly higher amounts.[95]

My story. . . .

Steve Dobson, General Manager for Product Support at JCB China

I spent a year on overseas voluntary work after graduation and was then employed as a Graduate Trainee by Coles Cranes and worked in their technical service department after I completed the 18-month training period. After three years with Coles Cranes I moved to work for JCB and have been with them ever since. During my 33 years with JCB I have held various positions working in Product Support, based in various locations in the UK and overseas – including Greece, India, Singapore and now China. Professional qualifications are important for this, because with certain government interactions they help your status. They also add credibility to your visa application, identifying you as a necessary person who can add value to local employees in terms of training.

What excites me about my work is that I am:

- working for a successful British manufacturing company who make a first class product of which I'm proud

- resolving customer concerns and exceeding their expectations

- ensuring that any problems that we report and are resolved will not be experienced by future users

- passing on skills in overseas locations which will be used after I've returned to the UK, enabling others to develop their knowledge.

I enjoy working abroad because of the necessity to make decisions quickly and independently of Head Office, which is far away, and that I am enabled to do so. Working internationally gives me more direct input into problem resolution and more influence over adapting our products to meet specific market applications in other countries. I also enjoy travelling, seeing different places and meeting people from a variety of countries and cultures.

There are some negatives of working internationally.

- The distance from family members.

- Difficulty in understanding the language, particularly true in China. This places you at the mercy of a translator which often leads to you not being given the full story, receiving only what the interpreter thinks you should hear.

- Dependent upon where you are based, uncertainty of some things we take for granted in the UK – for instance, clean water and reliable power were a major concern in India.

- The distress of constant contact with people who are underprivileged and need help, and being in a situation where it is not possible to help everyone.

There are also the intercultural challenges of:

- testing understanding where language is an issue: this is a major challenge

- trying to get the full facts on the cause of failures: you have to be systematic and patient

- recognising that what is said may not actually be what happened; for instance the operator may lose his job if he tells the truth about what he was doing when failure occurred

- accepting that 'yes' does not have the same literal meaning in some cultures and it may be what the person anticipates is what you want to hear; one needs to establish whether 'yes' means 'yes, yes', or 'yes, no'!

13

General and systems engineering

'A system is a set of parts which, when combined, have qualities that are not present in any of the parts themselves. Those qualities are the emergent properties of the system. Engineers are increasingly concerned with complex systems, in which the parts interact with each other and with the outside world in many ways — the relationships between the parts determine how the system behaves. Intuition rarely predicts the behaviour of novel complex systems. Their design has to iterate to converge on an acceptable solution.'[96]

'The idea of a system as "a set of parts which, when combined, have qualities that are not present in any of the parts themselves" is a very productive way of looking at the world — which turns out to be full of systems. Many engineered systems are much broader than the association with "engineering" might imply: the "elements" or "parts" of a system may include, for *example, people, processes, information, organisations and services, as well as software, hardware and complex products.* '97

Systems engineers integrate a breadth of engineering knowledge to inform their understanding of how an engineering system will perform given its component parts. Systems engineering describes a defined field of work and roles are easily identified in the jobs market. PEIs such as the Institute of Measurement and Control (InstMC), together with the International Council on Systems Engineering (INCOSE), promote the role of systems engineers. INCOSE also has a UK chapter (INCOSE UK) that forcefully argues for the importance of understanding a system fully:[98]

> *'It is not hard to know when system engineering fails, because when something important goes wrong it usually makes the news fast. People get killed, buildings fall down, companies go bust, the law becomes involved. But when system engineering goes right, no-one notices – which is just how it should be. The computer works when you switch it on, trains run on time, your flight lands on time and no one gets mad.'*

Systems engineers are required to think at a wider systems level rather than focusing on a specific area. They are therefore truly multidisciplinary in both their thinking and their interactions with others in an organisation or project. The role of systems engineers is to understand a problem so that they can inform the design solution. This process requires that the client's requirements are understood sufficiently well that the engineer can analyse the characteristics and behaviour of the system. Concerns and risks can be articulated early in the design process and action taken to manage them at a point when changes can be made with minimal impact on project cost and schedule.

The process of systems engineering may be summarised by the aide-memoire SIMILAR:

State the problem: what must be done, not how to do it

Investigate alternatives: extend or repeat as data becomes more certain

Model the system: create models for both the product and the process, use these to assess the alternative approaches and manage the project throughout its life cycle

Integrate: bring components together to work as a whole and consider how they interface with each other

Launch the system: create the system and test it

Assess performance: measure the performance against the expectations generated through the process of specification and design

Re-evaluate: observe the system's outputs and use this to improve performance by modifying the system, inputs, product or process.

In contrast, general engineering is often used as a 'catch-all' term for:

- a philosophical approach to engineering education that promotes a multidisciplinary understanding of engineering and which in turn is reflected in degree awards based upon a breadth of engineering knowledge

- degree courses that provide a broad base to engineering studies prior to specialisation

- engineering disciplines that do not fit elsewhere, for example in periodic assessments of the quality of university research output.

Of course the first two are not mutually exclusive and most courses in the UK are structured to promote both opportunities to some extent. General engineering is built upon the recognition that many engineering principles can be applied across the range of engineering disciplines, for example flow of electricity, heat and water. General engineering is not normally identified as a field of work in the same way as systems engineering, but many employers will list general engineering in their list of desired graduate degrees. The nature of that breadth of general engineering varies and will not necessarily adopt a systems engineering approach, but the similarities are such that the two disciplines have been brought together here.

The nature of the industry

The attractiveness to employers of general and systems engineers is based on the holistic approach that they apply to problems and the benefit this delivers. INCOSE notes that effective use of systems engineering can save 10–20% of a project budget. Such engineers are therefore sought after by employers who work with complicated systems, especially those that are safety critical such as aeroplanes, cars and nuclear power stations.

Typical employers for general and systems engineers include:

- Alstom (www.alstom.com): power and transport

- Atkins (www.atkinsglobal.co.uk): consultancy

- AWE (www.awe.co.uk): defence

- BAE Systems (www.baesystems.com): defence

- Jaguar Land Rover (www.jaguarlandrover.com): automotive

- Managed-Complexity (managed-complexity.com): system engineering consultancy

- Nissan (careersatnissan.co.uk): automotive manufacturer

- Raytheon UK (www.raytheon.co.uk): defence and security

- SA Capabilities (www.sacapabilities.co.uk): consultancy

- Siemens (www.siemens.co.uk): energy, industry, infrastructure and cities, and healthcare

- Thales (www.thalesgroup.com): aerospace, defence, security and transport.

Main sectors

Systems and general engineers find work in a wide range of industries and sectors because most have complex systems that must be engineered.

Defence

Here are some examples of defence-related projects that involve systems engineers in their development.

- Helmet mounted displays for pilots: in which enhanced data such as compass bearings, elevation, location of enemy aircraft and even the view below the aircraft are projected onto the helmet's visor according to the direction in which the pilot is looking.[99]

- Electric armour for protected patrol vehicles: armoured vehicles face a threat from shaped charge jets produced by rocket propelled grenades. Conventional armour designed to protect against these grenades will make the vehicles too heavy for many tactical situations. Electric armour uses a very high current to disrupt the shaped charge jet as it passes between two metal plates and thus reduces the weight of armour required.[100]

- Naval Electronic Warfare Operational Support Centre (NEWOSC): a shore based asset that provides naval vessels with a comprehensive knowledge of their environment and any electronic warfare threats. The system consists of an intelligence database of enemy weapons systems which it can use to assess the threat and issue mission-critical data to assets in-theatre.[101]

Other sectors

Other sectors that employ general and systems engineers are discussed in other chapters and include:

- aerospace (Chapter 9)

- automotive (Chapter 5)

- building services (Chapter 5)

- highways and transport (Chapter 6)

- energy and power (Chapters 5, 7, 8 and 11)

- rail (Chapter 6)

- water (Chapters 6 and 8).

Those wishing to gaining a deeper understanding of what constitutes a systems engineering project and the methodologies adopted by systems engineers should read the report by Dr Chris Elliott and Professor Peter Deasley.[96]

Entry routes

General engineering degree courses are available at a number of universities, whilst access to graduate careers in systems engineering requires an undergraduate degree in general or systems engineering or an aligned discipline such as:

- aerospace systems engineering

- automotive systems engineering

- chemical and bio-systems engineering

- communication systems

- computer systems engineering

- integrated mechanical and electrical engineering

- mechatronics

- systems and control engineering

- telecommunications systems engineering.

Graduate entry roles

General engineering graduates entered employment according to the following distribution:[3]

- 31% in manufacturing

- 30% in professional, scientific and technical activities

- 7% in construction

- 32% in other roles.

The INCOSE UK website includes job opportunities for system engineers.

For those entering general and systems engineering typical roles include the following.

- 'Bridging the gap' between different engineering disciplines involved on a project: for example software, mechanical design and electrical/ electronic hardware design.

- Evaluating customer or operational needs: to define the system performance requirements.

- Installing, configuring, integrating, testing, and documenting software packages.

- Interface management: to assure compatibility of all physical, functional and programme interfaces.

- Managing the customer requirements: to ensure the programme meets the customer's needs.

- Modelling system behaviour: how will the components work together? Optimisation of the system is not the same as optimisation of the individual components.

- Options analysis: researching alternative solutions or approaches to a problem or project.

- Researching new and emerging technologies: to identify if they might be used with new or existing systems.

- Reviewing designs: to ensure compliance with the project requirements.

- Risk identification and mitigation: predicting where a system might fail and acting to prevent or minimise the risk.

- System engineering analyses for factors such as:

  - affordability

  - human systems integration

  - maintainability

  - regulatory, certification and product assurance

  - reliability

  - safety

  - survivability

  - susceptibility

  - system security

  - testability

  - vulnerability.

- System life cycle design: planning for everything from conception to decommissioning of the product.

- System of systems integration: selecting and combining systems to meet a project's objective.

- Technical assurance: verification and validation of the system against specification and required performance.

Salary levels

General engineering graduates have a mean average salary of £29,673.[3] Based on a limited survey, systems engineering has an average advertised salary of approximately £39,000 with graduate positions in the £20,000+ range typical of other disciplines. Some 53% of advertised salaries were in the range £30,000–£50,000 and top salaries exceeded £70,000.

Part 3

Career development in engineering

14

Work experience, internships and other schemes to put you ahead of the competition

'Employers valued graduates with work experience as this demonstrated their motivation, interest, understanding and commercial awareness. Some employers insisted on applicants having work experience in engineering, whilst others felt that any type of work experience was useful as it provided graduates with an insight into work place values and practices.'[102]

'The evidence that a placement year improves employability opportunities is strong. Indeed, lack of work experience appears as a key barrier to young people, including graduates, in securing employment.'[27]

The National HE STEM Programme survey of unemployed graduate engineers clearly identified the importance of work experience for many employers. Of the unemployed graduates about a quarter had no experience and almost a half had work experience unrelated to their degree. In fact this was the greatest single reason reported for their predicament by the unemployed graduates in the survey. Many expressed regret that they had not availed themselves of a sandwich or vacation placement during their studies.

Employers identified the following benefits of work experience.

- **Engineering-related experience demonstrates both motivation and interest in an engineering career.**

- **It provides an understanding of the world of work.**

- **In some cases it develops commercial awareness.**

- **It helps to demonstrate employability skills such as those described in Chapter 4.**

An additional benefit of a year-long engineering placement for the student is the evidence that it leads to an enhanced degree classification.[103, 104]

Research has also shown that those graduates who undertook relevant work placements had the highest level of satisfaction that their graduate job is appropriate for them, whilst those with no work experience often felt that their job was inappropriate for them, perhaps because they are more likely to be in non-graduate jobs or unpaid work.[105]

This preference for graduates with work experience is supported by The Graduate Market in 2013[106] which reports that recruiters expect over a fifth of graduate engineering positions to be filled by those who have work experience, but more specifically have acquired that experience through internships, industrial placements or vacation work with their companies. The report reveals that over a half of the recruiters surveyed advise those with no work experience that they have little chance of gaining a place on their graduate training schemes.

The CBI[4] reported that 70% of employers would like students to better prepare themselves for the workplace, with 24% of employers noting concern about graduates' limited careers awareness and 42% stating that graduates should acquire more relevant work experience.

In the period 2003–09 about 13,000 engineering students undertook industrial experience each year, concentrated in mechanical, civil and electronic engineering.[107] With the advent of £9,000 tuition fees these numbers are expected to increase as students seek to offset the cost of study with earnings and universities seek to make their courses more distinctive.

Profile: University of York

The Department of Electronics at the University of York has been consistently ranked amongst the best electronics departments in the country for its teaching quality and world-leading research in electromagnetic compatibility, biologically inspired computing, music technology, wireless communications and nanotechnology.

It currently has over 70 full-time staff and over 500 students on a variety of programmes, specialising in subjects from avionics to nanotechnology.

Study

Since 2008 the National Student Survey has put York in the top three for student satisfaction in electronic engineering in the UK, as published in the *Sunday Times*. To complement the highly rated teaching quality, the department provides extensive laboratories along with specialist computing, nanotechnology, media and audio recording studio facilities. The degrees enable students to become the innovative engineers so keenly sought after by employers – or to pursue many other careers having gained a wealth of transferable skills.

All of the BEng/MEng degrees are fully accredited engineering courses, validated by the IET. This means they are suitable for graduates wanting to go on to become a chartered engineer. All degrees have a sandwich year option which can be taken between the second and third years of all our courses. For those on a four-year MEng degree, the sandwich year option can also be taken between the third and fourth years. MEng students can also choose to carry out their final year academic project within a company.

In the third year all MEng students participate in a major software project. This is organised in teams who operate as self-contained units. Teams can trade with each other and buy and sell software modules (with notional money!). The final product is a substantial working piece of software.

If students want to spend time further afield they can study on a year abroad or perhaps just take the opportunity to learn a different language alongside their course.

Courses

The range of courses includes:

- **Electronic Engineering:** This course provides a very wide range of knowledge and techniques in modern electronics.

- **Electronic and Communication Engineering:** The Electronic and Communication Engineering course gives students a strong electronics background with an emphasis on its application to communication technologies.

- **Electronic and Computer Engineering:** This is a Computer Systems Engineering programme combining the use of electronics and computer hardware/software.

- **Music Technology Systems:** This course focuses on the internal design and function of contemporary music technology systems within an electronic engineering programme.

- **Music Technology:** The Music Technology course is a creative music technology programme for students who are not taking mathematics at a higher level. The degree offered for Music Technology is BSc.

- **Electronic Engineering with Nanotechnology:** This course gives students a strong electronics background with an emphasis on its application to nanotechnologies.

- **Electronic Engineering with Business Management:** The Electronic Engineering with Business Management course comprises 35% business management, 65% electronics. It meets the needs of those with ambitions to progress to a management position.

- **Avionics:** This is an electronics degree with an emphasis on the design and application of equipment for the aerospace industries.

- **Foundation Year:** The Foundation Year is an entry route to all the courses for those who do not have relevant qualifications, particularly mature students.

Case study

Careers and industry

The placement and careers staff within the department provide advice on a range of summer placements, sandwich years and graduate jobs. Companies visit the department during the academic year to give talks on their careers opportunities and to talk to potential placement students. There are numerous close collaborations with industry ranging from basic research to product development, and a number of spin-out companies have been generated, often employing graduates from the department.

The department is an invited member of the UK Electronics Skills Foundation (UKFSF). This is a collaboration between major electronics companies and the public sector, working with universities to promote the electronics industry. It offers a range of scholarships that include annual bursaries, paid summer work placements and industrial mentoring and may include 12-month sandwich placements.

The Electronics Buildings and the Heslington Campus

The University of York was founded in 1963, and has three main campus sites, one in the historic centre of York, and two larger sites in the outskirts of the city at Heslington. The department is based at the Heslington West Campus. The central features of a landscaped garden have remained, and the university buildings line the winding banks of a lake, home to a large variety of wildfowl.

The Department of Electronics shares a building with the Department of Physics. The teaching tower is the tallest building on campus, with fine views over the lake. This houses the undergraduate teaching and project laboratories, as well as the audio studios and several of the department's computing labs.

Facilities and services

The department has a wide range of facilities used to support the teaching and research activities. Most of these facilities are also available in collaboration with industry, allowing direct input to project work.

- **BioWall:** an interactive development environment for bio-inspired computing.

- **Clean room:** facilities for fabrication and measurement of nano-scale devices, used by students from the first year of the nanotechnology courses.

- **Computing labs:** the department's own PC labs with specialised hardware and software to support the teaching.

- **Applied EM test facilities:** anechoic and reverberation chambers and open-area test-site facilities for researching electromagnetic problems from 10 kHz to 100 GHz.

- **FPGA and ARM-based development systems:** a range of tools for programming FPGA and ARM-based processors and system development.

- **Audio recording studios:** facilities for recording and producing acoustic and electronic music along with immersive laboratories.

- **Teaching laboratories:** for practical and project work and including iPad/iPhone workstations.

- **OPNET:** the department is a University Partner for OPNET: the industry-standard network simulation software.

Additionally, the department's Technical Support provides design and construction facilities for the department and industry, including PCB design and manufacture, digital manufacturing technology, 3D printing and a surface-mount assembly line.

Opportunities for gaining experience

One-half of all employers responding to The Graduate Market in 2013 survey[106] reported that they offer industrial placements of 6–12 months and/or paid vacation internships of three weeks or more for undergraduates. The report details the ratio of placements to expected graduate recruitment, giving an indication of their importance within each sector:

- consulting 42/100

- engineering and industrial 65/100

- IT and telecommunications 93/100

- oil and energy 80/100.

The average for all sectors reported was 67/100.

Employers offering placements and internships will often use them as a means for informally assessing students as potential future employees and if their experience of the student is positive they may offer a graduate position or rapidly advance the student through the initial stages of the formal selection process.

Types of experience

Placements

These are typically paid and of one-year duration. They may be integrated into a degree course (sandwich or intercalated year) or undertaken by agreement with the university. About a quarter of engineering degree courses formally offer a sandwich year,[107] but it is highly unlikely that any UK university would discourage experience of this type and many have units that actively promote such opportunities to their students.

Internships

Internships are typically between two and six months in duration and include vacation placements and placements in short-sandwich courses. For graduates whose lack of work experience is limiting their ability to land a job, a graduate internship may offer a solution. These have been developed in a number of employment sectors, including engineering.

Internships may be paid or unpaid. Some students may be prepared to undertake a short period of unpaid work experience where they believe this will give them a useful insight into the sector or company, but the practice is controversial where the internship leads to employment and restricts the opportunities of those who cannot afford to do unpaid work, or where the company has no intention of employing the interns but regards them as a source of cheap labour.

Internship opportunities are promoted through the normal commercial and university career services as well as through the websites of:

- The Graduate Talent Pool (graduatetalentpool.direct.gov.uk)

- Graduate Acceleration Programme in Northern Ireland (www.gapni.com)

- Talent Scotland (www.talentscotland.com)

- Go Wales (www.gowales.co.uk).

The Year in Industry Scheme (YINI)

Operated by the Engineering Development Trust, YINI offers paid work experience for a year immediately prior to university entrance or part way through the degree course. YINI provide support to the student through:

- a company-based mentor

- visits from the YINI team

- a free management leadership training course

- a course to maintain and develop their mathematical knowledge.

The scheme has 300 partner companies and about a quarter of participants subsequently secure sponsorship through university each year. As with other year-long placements there is evidence that the pre-university work experience leads to greater academic achievement.[108]

Volunteering

This includes engineering work through charities such as Engineers Without Borders. Whilst not a conventional workplace, it can provide many of the associated indicators and skills sought by employers such as a demonstrable commitment to engineering, international awareness, team-working and organisational skills, initiative and an ability to communicate with others. Graduates may find that offering their knowledge and skills on a voluntary basis may open up opportunities through employers observing their skills or the prospect to network with other potential recruiters.

15

Finding your first graduate role in engineering

'The competition amongst graduates to be employed by a "high brand" corporate is reflected by the competition between these employers to recruit the best talent available; it is a highly competitive process for both parties. The volume of graduate applications managed by a large recruiting company can be measured in thousands. . . . It is a rigorous risk-management process, reflecting the significant financial and human costs of making a wrong appointment.'[27]

Graduate employment prospects reflect the state of the economy and finding employment in an industry in periods of boom is much easier than in a recession. I recall an employer complaining to a meeting of academics that students were applying to her company in 'text speak' but then admitting that they still recruited them because they were 'desperate for graduate engineers'. In tougher economic times or when there is a high graduate/vacancy ratio a company can and will be much more selective in who they recruit. The good news is that engineering and industrial graduate vacancies have increased significantly over the last three years.[106] The demand from businesses for people with abilities at a graduate level, especially in science, technology, engineering and maths (STEM) skills is growing, particularly in high-value sectors such as low carbon, pharmaceuticals and digital media according to a CBI survey.[4] The same survey reported that over 40% of employers in the survey were experiencing difficulties in recruiting STEM skilled staff. There are therefore excellent employment prospects for engineering graduates offering the appropriate skillset.

Despite this demand for your skills, finding your first graduate job may loom over you and feel like a battle against the odds; but in reality it should have been a well-planned campaign from the outset of your studies. Not to the extent that you knew from day one which employer you wanted to work for on graduation (although you may have), but by reflecting, planning and acting to put in place the resources you will need to win your first graduate job. Your degree studies will have provided you with the technical know-how that an employer seeks, but that degree course and others like it have given the same opportunities to your competitors and if it was just a case of who achieved the highest grades from the best universities, graduate recruitment would be easy. Employers are looking for more than grades, they are looking for people who also fit the culture and needs of their company. Employers invest many thousands of pounds in recruiting a graduate and training them, and so they want to be reassured that their money is not just well spent but is also a good investment for the future.

Many of the published reports and surveys of employers and recruitment trends focus on the top corporate recruiters who have the most prominent presence in the graduate recruitment market place. However, significant numbers of graduates will find employment with small and medium sized enterprises (SMEs), such as specialist manufacturing units or consultancies, which will often use less formal selection methods. These opportunities, in what the Wilson Review of business–university collaboration termed the 'hidden job market', are easily overlooked and yet can offer rewarding career opportunities for those prepared to invest the effort and time into identifying them.

The key to success in finding a graduate job is: firstly to know who you are and what you want to achieve; secondly to know as much as possible about the sector and company and what they need; and thirdly to marry the two together to demonstrate that there is an excellent 'fit'.

facebook.com/BabcockGraduates

babcock

trusted to deliver™

Exciting. Dynamic. Fast-growing. That's Babcock

Where do you fit in?

Join our world-class graduate programme
and be a part of a fast-growing FTSE 100
business, working on prestigious multimillion-
pound projects in markets as diverse as power,
communications, rail, education and defence.
We have over 120 engineering, commercial,
finance and IT roles to offer top-quality
candidates who share our ambition and drive.
If that sounds like you, you'll fit right in.

See where you fit in at:
www.BabcockGraduates.com

Airports | Communications | Defence | Education | Emergency Services | Energy | Mining and Construction | Nuclear | Property | Rail | Training

Case study: Babcock International Group

trusted to deliver™

Jamie Legg

After graduating from Oxford Brookes University with a 2:1 honours degree in Business Studies and enjoying the summer to get over my university exploits, I embarked on the challenge of finding a graduate role.
Like most university leavers I signed up to the illustrious graduate websites and before long my inbox was inundated with job adverts, none of which actually cut the mustard. Until I stumbled across the graduate website of Babcock International Group.

Two years down the line I am now nearing the end of my graduate programme within the Education and Training business unit; and I can honestly say, Babcock is a fantastic place to work. I have been able to build up a wealth of experience whilst constantly being challenged and pushed within the roles I have undertaken; ranging from working in Business Development, interacting with high-profile clients such as EDF and British Gas, to leading projects on operational sites including organisations like Volkswagen N Group.

I am currently working on the London Fire Brigade training contract, where with my manager I am responsible for project managing the change in training delivery model. I am enjoying every minute of it.

Since day one I was surprised at how much responsibility I was given. Within seven months of being in the organisation I had gained my Advanced APMP certificate in Project Management, my first professional qualification, and I am hoping to add to this in the near future.

The professional training I have received has been excellent; workshops such as Team Working and Leadership have been fantastic and have also provided a platform in order to build networks across the business.

I feel very lucky to be part of a company that continues to grow from strength to strength in such an array of extremely interesting and vast sectors. If I was advising a fellow graduate about to embark on their career at Babcock, I'd say: There are so many opportunities, grasp what you can, but ensure you have an open mind. It is an exciting time to be an employee within Babcock International Group and I urge you to make the most of it.

Nadine Young

I studied Engineering with Product Design to a master's level at the University of Liverpool and, like most other students, I wasn't entirely sure what I wanted to do but

I knew I craved variety – which is why the Babcock Graduate Scheme was perfect for me.

If I tell you that my work so far has involved training to climb a 25m transmission tower to carry out a rescue, computer modelling transmission structures, taking leadership and presentation courses and meeting a member of the Royal Family – you'll see why Babcock appealed so much.

I am part of a two-year scheme that allows me to move to a different part of the business every six months. Throughout each placement I'm provided with a buddy and a mentor who introduce me into the particular division and ensure I get all the training I need throughout the six months I spend there.

My first placement was in the Power division and was based in Surrey. I spent my first six months as part of the Innovation and Development team researching, testing and implementing new technologies. I was trained in aspects of the job from computer modelling to the theory behind transmission structures and overhead lines.

For my next placement, I moved to a part of the Communications division where, after I had been sufficiently trained in tower climbing, I performed a tower rescue on a dummy! Site work so far has involved climbing masts in different parts of the UK to gather information required to complete jobs for clients. Getting to be hands on is the main benefit for me; it allows me to see and experience projects I'm working on and gain a deeper understanding of the division itself.

Babcock also provide graduates with a Professional Development Programme. I've had the chance to develop 'soft' skills, including technical presentation skills, teamwork and leadership and negotiation skills. The interaction with other graduates in different areas of the business definitely opens your eyes to the opportunities and variety that lie within Babcock and it's also a great networking opportunity.

Chartership has always been my core career goal and Babcock provide all the support I need to help get me to where I want to be. I have regular meetings with a qualified external mentor and am given the freedom to develop outside of the work environment. I get to experience four completely different areas of work, developing skills in each one. I can honestly say that I have loved every minute of my graduate scheme and I certainly haven't had the chance to get bored.

Who do you think you are?

The starting point in searching and applying for a graduate position is to understand yourself, to identify what you want to achieve, what sort of career you would like to develop, what your core values are and what you can offer to a potential employer. This will help you to target your applications at appropriate opportunities and employers with confidence. Many people find the process of self-promotion difficult; it runs counter to their culture and feels like crude bragging. There is often a sense that the recruiter should be able to 'read between the lines' and recognise your qualities, but the reality is that you will have to help them do this by providing a narrative to who you are and what you have achieved that flags up the skills they are looking for. Questions such as, 'what three words would your friends use to describe you?' are designed to sidestep people's natural reserve and allow them to speak more freely about their qualities. As a part of your self-reflection why not ask your family and friends how they would describe your strengths?

What are your assets?

The first step is to give consideration to all the things that you have accomplished that define the package of knowledge and skills that you will offer to a potential employer.

- Academic studies: what subjects have you studied and what projects have you undertaken? What have you learnt and achieved in the process?

- Achievements: what have you done that has achieved the recognition of others as something well done? Have you been elected as a committee member or won a prize? Have you raised funds for some worthwhile activity or cause?

- Hobbies and pastimes: how do you spend your time? What activities or subjects really engage you? Is there something you would do all day, every day if you could make a living at it? Which student societies have you been involved with?

- The differentiator: what is it that differentiates you from your peers? Is there an example that you can use to make your application or interview stand out from the rest?

- Work experience: what experience have you gained irrespective of the relevance of the work to your chosen career? Can you articulate the skills you have acquired? The survey by High Fliers Research Ltd[106] reported that graduates without previous work experience are unlikely to be successful during the selection process and have little or no chance of receiving a job offer for graduate programmes in over a half of the companies surveyed. The nature and value of work experience are discussed in Chapter 14.

- Employability skills: how can you demonstrate the soft skills that are important to employers? Employability skills are the most important factor taken into account when businesses recruit graduates, with 82% of employers valuing them.[4] The CBI and National Union of Students[109] identify employability skills as:

  - self-management: your readiness to accept responsibility, flexibility, resilience, self-starting, appropriate assertiveness, time management, readiness to improve own performance based on feedback and reflective learning

  - team-working: respecting others, cooperating, negotiating, persuading, contributing to discussions, your awareness of interdependence with others

  - business and customer awareness: your basic understanding of the key drivers for business success and the importance of providing customer satisfaction and building customer loyalty

  - problem-solving: analysing facts and circumstances to determine the cause of a problem and identifying and selecting appropriate solutions

  - communication: your application of literacy, ability to produce clear, structured written work, and oral aptitude, including listening and questioning skills

  - application of numeracy: manipulation of numbers, general mathematical awareness and its application in practical contexts (e.g. estimating, applying formulae and spotting likely rogue figures)

  - application of information technology: basic IT skills, including familiarity with commonly used programs.

What are your aspirations?

Think in terms of a series of time frames: what would you like to have achieved in 5, 15 and 25 years? Here are some questions you might ask yourself.

- What type of role would I like to hold?

- What sort of life/work balance do I prefer? Will that change with time, for example if I have a family?

- Do I want to travel? If so, how frequently? For example a job might involve periods of international travel on a weekly basis or you might wish to work in jobs located one place but move locations as you change jobs.

- Am I driven by challenge, money, esteem and/or a sense of doing good?

Having determined your aspirations consider the extent to which the assets you identified match your ambition. Are your aspirations realistic? Is there any further training you need to undertake to achieve your ambition, for example if your ambition is to become a leading researcher and/or professor in your chosen field you will need to plan to complete a research qualification such as a PhD or engineering doctorate (EngD)?

In life all manner of things can cause your career plan to change and some people are more opportunistic than proactive in developing their career, but at this initial stage of winning the first graduate role employers will expect candidates to have some idea of what they want to achieve.

Hopefully this will not be the first occasion on which you have given consideration to your assets and aspirations, because these are useful reflections and motivators during your studies and should have formed a part of your ongoing personal development plan (PDP). A PDP provides a supported, structured, dynamic and reflective record of your achievements and aspirations that is designed to assist you in targeting your activities towards achieving your goals. The Quality Assurance Agency notes that:

'PDP results in enhanced self-awareness of strengths and weaknesses and directions for change. The process helps learners understand the value added through learning that is above and beyond attainment in the subjects they have studied. Crucially, it relates to their development as a whole person, but also has benefits for others with whom the learner interacts.'[110]

Do your research

Feedback from a number of engineering recruiters for large companies complained that increasingly they are meeting with students fixated on 'how can I get in' rather than trying to understand the company or the job role and career. They also observe applicants tending to take a scattergun approach to applications – applying for areas of engineering they are not actually interested in. A targeted and properly researched application will always fare better than a generic application used for several potential employers, but focused on none of them. More important than this is that the research stage will help you decide if the job/employer offers the type of career move you are looking for. It may also reveal information about the financial position of the company and whether working for them offers a secure future. Engineering students will normally have studied how to interpret business metrics and they should use this knowledge to their advantage. It is also wise to think broadly about the company. You may be applying for a position in one section of the organisation, but an awareness of other divisions within the company will show that you have done your 'homework' and not just relied on a quick scan of their website. At the end of this process you should have a firm grasp on the work that the company does and what about it is attractive to you and why.

Information can be acquired from several sources.

The job description

The job description will list the specific criteria that the employer wants. Sometimes for graduate roles these will include specific technical knowledge, but frequently they simply ask for a relevant degree with a minimum classification and a list of employability or soft skills. Where the description is non-specific this may reflect a situation that the company has a number of vacancies or has a graduate training scheme that will distribute the successful candidates across several functions.

Where the job description does specify particular requirements you should ensure that your application highlights how you meet these. Employability skills are communicated more effectively through carefully crafted examples of where you have used them rather than as an unsubstantiated list of claimed skills. For example team-working skills may have been demonstrated in coursework or group projects where you worked with other students, or they may have been demonstrated through a work placement or student society. The important thing to remember is that you must focus on what you contributed to the success of the group and not just what the group achieved as a whole. Some students find it difficult not to share the credit with the team as a whole, but the employer may interpret this generalisation as an indication that you were a passenger in the group rather than an active participant.

Company website

When researching a company via its website you should focus on the content rather than the presentation and avoid being dazzled by an expensive animated interface or stunning graphics. You should search for the following sources of information.

- **Annual reports:** these will give you an insight into the recent developments and financial well-being of the company. Often you will be able to access more than one year of annual reports and you may find it useful to compare two or three recent ones to check on how representative the last reported year's activities are compared to the company's recent past.

- **Specialist reports:** what is important to you in a prospective employer? Look for reports that deal with that aspect of the company's philosophy and policies. For example you may find reports on corporate responsibility, sustainability, health and safety, etc. What do these reports reveal about the company? Think about what they have included and what they have omitted; how do their reports differ from their competitors' equivalent reports? What are their core values? Think of your task as a piece of comparative coursework in which you identify strengths and weaknesses in the company's public profile. Are there any good observations or questions that this research suggests for you to use if you are fortunate enough to progress to interview?

- **Case studies:** these will give you an insight into the company's most interesting and successful projects. You may also gain an understanding of the company's client list and technical capabilities.

- **Graduate profiles:** will provide information on their past recruitment strategies and provide insights to the graduate training programme.

Internet

The company's own website exists to promote their interests, but what do other sites say about the company? Do they have any negative reports, for example from customers? Have there been any significant health and safety incidents involving the company?

Depending on the size of the company you may be able to pick out relevant names of key leaders and check for information on their biographies through sites like LinkedIn. Use that information to search for papers, talks and PowerPoint presentations that give you an awareness of their thinking. Websites such as glassdoor (www.glassdoor.co.uk) can provide an insider's view of the company, but beware of basing your decision on the references of a small group of people.

Competitors

Viewing competitor company web pages can sometimes give a useful comparison with which to critically appraise what the company says about itself. Knowledge gained from competitor sites can also be a useful means to demonstrate broader sector awareness or to highlight positive features of the company at interview.

Sector publications

Many sectors of industry have dedicated trade magazines or specialist interest groups that provide current news stories within the industry. If you are targeting a particular industry you should be viewing these publications, many of which will have an online presence. Interviewers will expect candidates to know something about the present state of the sector, what sort of projects are current and what is forecast. You may also pick up useful information about industry awards, positioning or other facts about a company that you can use in your application or interview to demonstrate that you are interested and informed about the company and industry.

Alumni

Some universities have schemes for putting applicants in touch with past graduates who work for target companies and some companies actively use recent graduates to target students from their preferred university courses. Use these contacts to ask questions about the company, values, recruitment process and so on. This should not be considered a shortcut to avoid the research that you should be doing using the other sources discussed here. If you cannot find a contact ask your friends and acquaintances if they know anybody working for the company and can give you an introduction.

Application process

If you have analysed your assets and aspirations and fully researched the company and the graduate role under consideration and have established that they match, you are now in a position to apply. The next steps depend upon the application process dictated by the company. For large or popular graduate destinations this will often have multiple stages such as:

- a web-based application form

- an initial assessment

- a telephone interview

- an assessment centre

- further interviews.

There are no hard and fast rules about the number of steps or the form that they take. Some companies may have a recruitment process focused only on submission of a CV and interviews. If you are applying to a company that includes multiple stages, any of which may reject your application, you will need to apply different strategies appropriate to the stage. A software robot will be programmed to search for key evidence, such as type and class of degree, and will execute its task impersonally. You will either fit the criteria or not! Once you break through to the interview stages you are dealing with people and have an opportunity to show your skills and interest in their best light. The individual setting out to win their first graduate position must therefore piece together the evidence and stories that reveal their skills and character to potential employers, such as describing their involvement in a project, work placement or student activity.

In less regimented processes you will be invited to submit a CV or may do so as a speculative application. If you are submitting a CV it should always be accompanied by a covering letter. You should never send a CV and covering letter if the advert specifically asks you to apply using the company's application form. If the form is a paper submission it is appropriate to attach a covering letter unless specifically instructed not to do so.

CV

Your CV has one purpose: to provide, in an easy to absorb format, sufficient information for the recruiter to progress your application to the next stage based on a 20–30 second appraisal of it. In that time they need to ascertain that you have the skills and background that they are looking for. Your CV is therefore a marketing tool setting out how well suited you are to a role by providing evidence in an accessible form. The content of the CV should be influenced by the information required in the job description or advertisement. Ask yourself the question, 'What are they looking for?' and give it to them. For example if the graduate role is focused on a particular area such as stress analysis you should highlight any relevant modules, projects or experience that demonstrate you have the knowledge and skills desired.

A CV should be no longer or shorter than two pages, avoid long blocks of text and be tailored to the employer and job advertised. Here is some of the information the employer will expect.

- Your contact details.

- A succinct educational history focused on your degree. What degree have you read and what did it involve? Give results in terms of class (First, 2.i, 2.ii, etc.) rather than as a percentage. Help the employer see the relevance of your degree to their needs.

- Your work experience: what have you done and what relevance does it have?

- Your skills: more than just languages and IT, describe your broader transferable skills and where you have exercised them.

- Your achievements: more than academic grades. What makes you special or different compared to other applicants? This might include voluntary work, being an office holder who has organised a major event or an award.

- Your interests and extracurricular activities: what excites you? Perhaps you have a hobby that is relevant to the job being applied for, for example a civil engineering graduate with climbing and abseiling skills will be of interest to companies specialising in working at height such as abseiling down tall structures to inspect their structural integrity.

- Contact details for your referees. For a new graduate these will normally include one or two academic staff who know you well, such as your personal tutor and project/dissertation supervisor. Those with work experience may include a previous manager.

Covering letters

A covering letter is not always required but where it is it should be focused, neat and professional in approach. The covering letter may well determine the recruiter's first impression of your application and thus you, so ensure it will work in your favour and not against you. If submitting the letter electronically the following criteria should be adapted accordingly.

- Paper: one side of A4, white or cream in colour.

- Style: use the same paper, font and font size used for your CV. The two documents should immediately show that they originate from the same 'stable'.

- Content:

  o The company name and address. If you know the name of the recipient address the letter to them adopting a formal style.

  o The post applied for and the job reference number.

  o Where the post was first seen advertised.

  o A succinct explanation of your motivation for applying to their company.

  o An explanation of your suitability for the position including any relevant work experience.

  o Round off the letter on a polite but optimistic note and sign off formally using 'Yours sincerely' if you know the name of the recipient and 'Yours faithfully' if you do not know their name.

The letter should convey your enthusiasm and demonstrate that you are applying from a sound understanding of the company and the role, gleaned from your research. You should ensure that the content is relevant, interesting and original to the extent that you are not simply echoing their marketing material. Do not reproduce the material in your CV, but make reference to it as needed. Wherever possible use evidence in the form of examples where you used a skill or performed a role. This information might relate to your work experience, engineering education, extracurricular activities and voluntary work or life experiences in general. Finally make sure that you have spell-checked and carefully proofread your letter. If possible review the letter after a night's sleep to avoid reading what you think you have written as opposed to what is in the document.

If you are emailing your letter and CV to the company be careful not to fall into the casual style that you normally use for emails. You should construct the email as a professional document as though you were writing a letter for the post and send it from your university email account.

Application forms

When completing application forms the required information is clearly set out for you, but you should approach the task of filling it in with the same care and professionalism you would apply to crafting your CV. Complete all of the required sections and ensure that your facts are accurate even if you think the uncertainty in your answer is unimportant. Failure to do this may give your future employer grounds to withdraw their offer or terminate your employment.

Structure answers to biographical questions such as those exploring educational or employment history so that you present your most recent situation first. In describing your responsibilities, use bullet points and commence each one with a verb because this will emphasise both your active contribution and achievements. For example there is little remarkable about being a member of the Rugby Society, but taking responsibility for planning and organising an international team tour shows imagination, organisational, team and financial skills. Example use of verbs is given below:

Employment history:

Team Leader, Student Union Café November 2011–Present

- **Assisted** in drawing up team rota . . .

- **Ensured** all sections checked and re-stocked . . .

- **Proposed** alternative marketing to increase sales . . .

- **Co-ordinated** team training . . .

- **Won** best SU Outlet Prize and received £250.

Activities:

Activity Coordinator, University Volunteers October 2012–July 2013

- **Liaised** with Social Services to identify children needing support . . .

- **Planned and organised** weekly events for children . . .

- **Communicated** with parents/guardians . . .

- **Drove** minibus to collect and return children from/to their homes . . .

- **Co-ordinated** support team from other volunteers . . .

If the form has an open section for additional information use it to highlight how you have demonstrated the skills and knowledge that the company are looking for. Do this by **CAR**:

C – describe the **context** of your story and what your objective was

A – outline the **action** you took to achieve it

R – summarise the **results** of your actions, focusing on the impact you made and what you learnt from it.

For example one of my students produced software to improve the distribution of online teaching materials to students within my department. In this case he can describe the **context** in terms of his experience as a student and the frustrations that he and other students faced in terms of their user experience when accessing support materials from the departmental website. His **actions** were to raise the issue with me in my role within the management of the department; persuade me of the need to improve the system and provide a proposal and budget for his proposed solution. Having achieved permission and funding he recruited a small team of students who between them delivered the project. He was able to summarise the **results** in terms of the product, the 600 students who downloaded it upon release and the university prize he won for this and other contributions. In the process he reveals an entrepreneurial flair and skills in negotiation, project planning, team building, software design and authoring, marketing and working to a deadline.

Managing your social media

The ubiquitous social media channels provide opportunities either to enhance or destroy your recruitment opportunities. A High Fliers Research Ltd report reveals that a significant majority of recruiters surveyed reported an increased emphasis on the use of social media to promote employment opportunities.[106] Social media platforms are designed to promote networking and it is not uncommon for our personal networks to also generate employment leads through individual contacts. As a student there are many opportunities to build up a network of contacts that may prove useful in the search for your first graduate job. Some of them are listed below.

- Friendships with students who are further through their course of study than you. The student one or two years ahead of you will graduate before you and may well be working in the sector you are interested in. They will be able to provide feedback on useful information about their employer, current opportunities and the recruitment process.

- Your university or department may have social media accounts for linking with alumni that you can join.

- Industrial visiting speakers, mentors and members of course advisory panels may be willing to permit you to link to their profile on LinkedIn.

- Friends of your family may work or have contacts in your chosen sector.

- Academic staff may have links with contacts in industry and past graduates.

Be thoughtful and strategic in who you approach; this is not a competition for who has the largest network. Not everybody will be willing to share their contacts freely with students, but a polite request or invitation to connect costs nothing and may prove useful if successful. Having created your network you can use free services like that offered by the online advertiser adzuna.co.uk to flag up the links between members of your network and the vacancies they list.

Facebook

Often a Facebook account will have been created as a means to link with family and friends and will contain references to your lifestyle that you would never dream of including in a CV or covering letter. Stories of events, derogatory comments on individuals or companies and those unflattering end-of-party photographs may well deter potential employers who decide to search for your profile. In some careers, for example teaching, you will not want others, e.g. pupils, to have access to details about your personal life through Facebook, and whoever your employer is you should avoid making negative comments about them or your colleagues. If you are not sure that your profile will reassure a potential employer, it is wise to take action to restrict access to it.

If you believe that your Facebook entry is suitable for viewing by potential employers you can use it to your advantage. It may not be the most obvious vehicle for job searching but flagging up the type of work you are looking for will tap into your existing network of friends and their contacts. You can also access numerous Facebook profiles for companies or job advertising/seeking groups.

LinkedIn

LinkedIn is a network for professionals and companies. Unlike Facebook there is little danger of confusing your private and professional life in your LinkedIn profile. It is therefore a much safer tool for promoting yourself to potential employers and researching them. Your approach should maintain a professional style when presenting information within your profile. LinkedIn also supports groups that you might join and contribute to raise your profile, but think carefully about how your posts will be read by potential employers.

Twitter

As with other social media platforms, Twitter provides a means to connect with and follow those companies that you may wish to work for. Twitter also carries notifications of job opportunities and a quick and simple way of reviewing the people and companies that have posted these is to direct Google to search the Twitter server for key words using the #jobs tag, e.g. 'site:twitter.com systems engineer #jobs uk' will list systems engineering vacancies in the UK.

Blogs

If you have a blog the same warnings apply as for the other social media platforms discussed here. If the material is controversial it is unwise to draw it to the attention of potential employers.

Assessment centres

Assessment centres provide an opportunity for recruiters to use situational judgement tests and assessment using 'real life' scenarios. Situational judgement tests are a form of psychological aptitude test that uses multiple choice answers to questions describing typical workplace scenarios. These provide a more accurate picture of the candidate's fit for the role and their potential to perform well in it. The process also provides the candidate with a reality check; will they really enjoy working in this type of role? Assessment centres allow recruiters to observe a candidate's ability to:

- work in teams with others

- process and use information

- demonstrate skills such as leadership and initiative

- complete day-to-day tasks.

The precise nature of an assessment centre will depend upon the company running it, but the type of activity that candidates are asked to engage in might include reviewing technical and financial data, weighing different options and making recommendations for future strategy, investment or equipment purchase. The scenario may be carefully crafted to identify how well you have understood the company, its work and existing strategy. A recommendation that is logically sound but cuts across the company's values without acknowledgment may well be the 'wrong' answer.

The assessors will be less concerned with what you know and more with how you think as you are presented with situations or problems and respond to them. Before attending an assessment centre you should refresh your knowledge about the company and your relevant skillset. Arrive early and take the opportunity to settle into the environment and engage with other candidates. This will help you relax and your interactions with other candidates will help when you are asked to work together in groups later. Remember that you are being judged against a set of criteria and not against the other participants; it is therefore important that the assessors know who you are at all stages and something as simple as ensuring your name badge is easily visible throughout the day is important.

Be attentive to what is happening at all stages because the assessment centre will have been carefully planned with purpose throughout. Information sessions about the company may well provide important information that should influence decision-based activities later in the day, when you will wish to align and justify recommendations you make with company strategy. For example arguing that a food manufacturer noted for their vegetarian ethos should invest in meat processing is unlikely to be successful.

Interviews

What will the interview be like? This a pressing concern for any graduate, especially if it is your first interview. Even if you have taken the opportunity to have a practice interview with a university careers adviser, or an employer who has visited your faculty to offer students some interview practice, there may be considerable differences between either of these experiences and the real thing. This section will highlight some key areas that characterise the way graduate interviews for engineers (and by engineers) are conducted. The nature of the interview may vary with the nature and size of the company, for example SMEs may have a more informal approach to the interview than large organisations.

Generally speaking, today's job interviews are no longer left to the discretion of the manager concerned to get to know the candidate by having a chat about their CV. In a job interview, as in the assessment centre, you are being judged against a set of criteria. Let's assume that you already know that a set of competencies chosen by the company's HR department will dictate the particular set of questions you'll be asked at interview, and that each of your answers will be scored against a corresponding set of **behavioural indicators** for the competency concerned (an example is provided below). This is a systematic process that may very well appeal to you, as an engineer. The structured nature of the so-called 'behavioural' interview arose through the need for employers to be accountable for their decisions about who should fail or succeed. The aim is to reduce the potential for 'interviewer bias'. In other words, if they ask each candidate the same questions and score everyone on the same criteria, their decisions can be understood to be objective and analytical.

So, although 'Tell me a bit about yourself' is thought to be a popular question in job interviews, and one that you are very likely to be asked by a careers adviser or the director of a SME, an experienced engineering recruiter will tend to avoid this question. One engineering recruiter observed, 'By using consistent approaches, and without asking them to talk about themselves off their own bat or off their own pre-prepared script, we can get down to business pretty quickly.'

When this engineering recruiter talks about 'consistent approaches' he is referring to the fact that he asks mainly behavioural questions. These questions require you to describe an actual experience that you have had, which your interviewer evaluates by counting how many of the behavioural indicators that demonstrate a particular competency actually occur in your story. These will enable each competency to be given a rating, usually out of 5. See below for an example for the behavioural indicators interviewers look for in order to recognise whether a candidate demonstrates desired competency:

Competency: Teamwork

Behavioural indicators

1.   Co-operates to meet team goals even at expense of personal preferences.

2.   Addresses conflicts or issues within the team in a positive and open manner.

3.   Uses understanding of different interests and agendas to achieve positive outcomes.

The candidate's answer can be scored highly if the interviewer can tick the boxes for all of the behavioural indicators.

Teamwork

Teamwork is a good example because it is both a key competency for engineers and the make-or-break question in many engineering recruitment interviews. In other words, with this question, things can go very right — or very wrong.

Your experience of working in a team, and how you talk about it, may be crucial to the outcome of the interview. Your best example of teamwork could be from a job or from voluntary work or from an outside activity you have been a part of, such as a sport: all of these are potentially very useful. At the same time, you will need to be prepared to give an account of a project undertaken on your engineering degree, so you should understand that whatever you say about this experience will inevitably be assessed as an example of teamwork. It is important to be aware of the various opportunities this offers to present your abilities favourably to the employer.

It helps to understand first of all that an effective teamwork story in an interview is not necessarily a story about an effective team. It is an effective **story** that is **about** working in a team. Recruiters want to hear you talk about your team members — they infer from this how you will relate to colleagues in the workplace. And they will almost certainly want to hear you talk about a difficulty or a conflict in a team, and how you approached it. How did you interact with others to come to a decision, solve a problem, deal with challenges, and so on? What did you learn from that? It does not matter if it did not work out that well, provided that you can identify what you learnt.

In fact, graduates who are happy to talk to a recruiter about difficulties and conflicts in a team, and can also do this in a sufficiently neutral manner (i.e. not blaming their team members), are more likely to succeed in job interviews. The opposite is also true, in the sense that those who deny there were any problems in their team are effectively unable to show their analytical abilities in discussing a conflict objectively. Given that most student teams are dysfunctional and have one or more members who are not sufficiently motivated or do not pull their weight, and this is well known to any engineer who might be interviewing you, it is even more important that you can talk openly about the way the project went but without seeming to allocate blame. An engineering recruiter advises: 'I would absolutely counsel you to have . . . a set in your mind of examples of conflict; of leadership; of teamwork; and things like that, because they're very hard to think of on the spot.'

In order to help explain what is right about a teamwork story, it's useful to understand what else can go wrong.

And what was your role in that team?

When the interviewer asks about your project or other teamwork experience, unless you can identify the actions you personally took in the team endeavour, you run the risk of giving the impression that you were a passenger and not an actively contributing member of the team. Be careful, however, not to give the opposite impression – that you did all the work yourself, with a few hitch-hikers along for the ride. You need to talk about what you personally did in the team as well as what the team achieved collectively. This means being aware of maintaining a balance between the number of times you say 'I' and the number of times you say '**we**' in your story. One useful tip is to try and describe how you and the individual members of your team interacted with each other to get things done. Because you are reporting interactions, you will automatically be incorporating your own actions within the actions of the team. It is very effective if you include what people said to each other during these interactions. Here is an example: versions (a) and (b) both include what people said to each other as they interacted. But which version is the best?

(a)

'So he asked me if I could change the parameters on the control urgently and I told him I'd prioritise it and get back to him later the same morning. So although we had to work separately for a while, the project came in on time and the client emailed me to say they'd recommended me for an internship as a result.'

(b)

'So he said "Can you change the parameters on that control ASAP?" and I said, "Sure, I can prioritise that and get back to you this morning." We had to work separately for a while, but the project came in on time and I got an email from the client saying "We're recommending you for an internship".'

In a job interview, (b) will always work better than (a). This example clearly illustrates the difference between telling a story and writing one down, as you would do on an application form.

What strengths would you bring to my team?

Another typical pitfall identified by researchers who have analysed engineering graduate interviews is that engineering recruiters may ask other questions with the word 'team' in them that are not about teamwork itself. 'What strengths would you bring to my team?' is easily mistaken for a question about teamwork skills. It is safest to assume that when the word 'strengths' is used, the recruiter means something engineering-related, but if in doubt ask for clarification from the recruiter as to whether the question is about team-working skills or technical skills.

Not all questions about teamwork are behavioural. Other types of question, such as strengths and weaknesses, are less frequent, but require preparing for. Many students worry about these questions more than any others, trying to think of a 'good' weakness (a weakness that is also a strength; a weakness that you are still working on, etc.). Note that it is almost more important how you frame the strengths or weaknesses, than the strengths or weaknesses themselves. For example you may have expected everybody to grasp ideas as quickly as yourself, but having found otherwise you have adapted your speed and style of explanation, perhaps using sketches to illustrate key points more clearly. Remember when you are preparing your answers that they may reveal more about how effectively you communicate than they do about your strengths or weaknesses. All employers are looking for communication skills; engineers are expected to be able to explain things well; this, not surprisingly, includes the way you communicate in the interview itself.

How would the members of that team describe you?

Consider another type of question about teamwork whose underlying purpose is to get you to identify your strengths (and weaknesses) **indirectly**. This is a question intended to sidestep your natural reserve and allow you to speak more freely about your qualities. It is an indirectly framed question, so you can take advantage of the opportunity to offer an indirectly framed answer. As a guide, your answer should include some positive things and one thing that you could improve on.

Leadership

Leadership is a competency separate from teamwork, but leaders and teams are inseparable (even if there was no **appointed** leader in your team). If you are being recruited for a graduate rotation programme you will very likely be asked a leadership question as well as a teamwork question. When you answer the leadership question, you will be talking about other people, and you will probably have to talk about a people-related problem. So you should follow the same rules for talking about your team members as for a teamwork question. For example you may have led a team on a project where one of the members was unable to complete a task on schedule (this may have been because they had different priorities, were ill, lazy or faced a convergence of deadlines for coursework); your answer could highlight the response from the team as a whole and your actions in particular, describing how the team supported the individual and adapted to the delay.

There are various other behavioural questions you may be asked in an interview apart from leadership and teamwork. For engineers, innovation, problem-solving (which are related), and time/stress management would be the most likely. Similarly, you will need to prepare a story based on a real experience to demonstrate your competency in each of those areas.

Technical questions

You may be asked a technical question in your interview. The underlying question here is whether you are good at explaining things in a simple and logical way. People who are good at explaining things very often use analogies to get their point across. Engineers are people who typically use analogies to represent complex processes. This is also a key feature of the way successful leaders in any field communicate their ideas. Analogies are essentially metaphors and these are common in every language and culture – a universal tool of communication.

Candidates faced with technical questions often struggle to recall details of equations and formulas, when this is not what the interviewer requires. It is quite obvious when a candidate is searching their memory for what they read in a textbook, when a common-sense answer is actually what the interviewer is trying to elicit. An ideal answer is an explanation that would be understood by a layperson, and an analogy is a very good way to do this. For example, if asked 'What is voltage?', the respondent may reply, 'current times resistance' (or 'IR') or they may describe it using the more effective approach of a water pressure analogy.

Orientation to engineering

'Orientation to engineering' is a category of questions that evaluate your reasons for wanting to be an engineer. You will need to explain your choice and recruiters will decide whether or not your explanation is relevant. This relates to a wider concept in recruitment known as 'person–job fit'.

Questions such as 'Why did you want to be an engineer?' and 'Why did you get involved in engineering?' all require you to demonstrate your interest in actually doing engineering, not just through talking about your project, but also as it relates to what originally Inspired you and what you do outside of your studies. To quote one recruiter, does all this add up to 'a coherent picture of an engineer'?

'Passion for engineering' is a phrase often used by recruiters. Even if passion may seem rather an extreme emotion to express about engineering, your main concern is how to demonstrate it. **Stating** that you are passionate about engineering is not the same thing as **demonstrating** that you are passionate about it. What kind of evidence is considered to demonstrate a passion for engineering? Note that this is not an issue of personality. You don't need to be an extrovert to demonstrate passion for engineering: employers say they know how to recognise it in introverts too. Are you interested in your own project? If you can explain why you are interested in it to someone else, you will have succeeded simultaneously in demonstrating good communication skills.

Why did you get involved in engineering?

Reasons for choosing engineering that recruiters consider relevant fall into two main areas. One is a formative childhood experience in which an adult, usually an older relative, inspired you to participate in an engineering-related activity. This would typically involve some kind of hands-on tinkering with mechanical or electronic components and/or tools. This person could be said to be a role model. Note that many engineering graduates come from engineering families and as such have role models, but merely stating that you were inspired by a particular role model is not considered evidence of passion for engineering. You need to explain an actual situation in which that person inspired you and what technical thing it was that you got involved in doing.

The second area that recruiters consider relevant is related to outside interests: in other words, you were doing something **technical** in your spare time that you were passionate about, which led you to choose engineering as a career. What was it and why did you enjoy it?

When you are asked why you chose engineering, if you haven't got an answer corresponding to one or the other areas above, how should you respond? Remember that you are providing evidence of passion for engineering. So 'because I was good at maths' is not a satisfactory response, because it does not reflect what it is about engineering that appeals to you. Can you explain what it was about maths that you enjoyed and, for example, how being good at maths helped you solve a real-life problem?

Outside interests

'What do you do outside of all this?' is an engineering recruiter's way of sidestepping the words 'hobbies' and 'pastimes'. A popular belief about CVs is that the interests you list should reflect an all-round person. Research suggests that this is not the case if you are an engineer. What you do outside study should primarily reflect a person who is 'really into engineering' and a person who can work well in a team. The first is demonstrated through activities outside your studies that are engineering-related, preferably involving hands-on tinkering of some kind; societies such as Engineers Without Borders or those linked to a PEI are good examples. The second is assumed by recruiters to be demonstrated through participation in team sports, but may also be demonstrated through other team focused societies and roles, as one engineering recruiter observed: 'Our business is about teams – about people with certain skills working with other people with complementary skills ... So that's where the team sport you choose is interesting: we get a bit nervous about golfers and tennis players; we get even more nervous about people who like reading books.'

The engineering recruiter quoted here touches in quite a light-hearted way on the widespread assumption that what people do in their spare time reveals crucial information about how they relate to others. However, you should not underestimate the significance of the type of sport a candidate plays.

The first part of the chapter noted that job application forms are sometimes initially evaluated in an automated process. One particular feature of this practice that has been reported to occur is that forms and CVs are pre-screened for team-oriented or solo pursuits. The 'wrong' sports may preclude a candidate from being shortlisted for interview. The 'right' sports, however, can boost a candidate's chances in the interview itself, especially as they may offer an opportunity for establishing some common ground with the interviewer.

To summarise, what you do in your spare time is required to signal 'engineering' and this can be communicated to a recruiter both directly through evidence of hands-on tinkering activities (a technical hobby) and indirectly through evidence of involvement in team sports.

Talking about your project

Finally, what about the evidence of passion for engineering you can provide when talking about your project? Like any teamwork story, when you talk about your project you should set the scene by describing what your team set out to do, foreground your own actions as you report how it evolved, explain how it worked out, and preferably conclude with what you learnt from it. Going into sufficient detail about the technical process driving the idea behind your project shows you understand it and know how to communicate it. This will reflect your potential to successfully communicate an idea to a client as well as to your peers and so is very important.

As we have previously noted that this is not an issue of personality but one of communication. In the normal course of events a student who has actively engaged in group project work will have no difficulty in describing what they did and achieved with exuberance. You might usefully consider if there is a simple visual aid that you might use such as a poster or photograph linked to the project.

Person–job fit and person–organisation fit

Bear in mind that your overall goal is to show employers that you fit, and this includes knowing how the company functions and exactly what it does. But also bear in mind that while your interviewers genuinely hope to establish that you do fit, their overall goal is actually to find out if you do not fit. It is important to find out enough about the company to help you prepare for the interview. You can then draw on appropriate experiences in your answers to demonstrate to them you are a good fit for the role and a good fit for the team, as well as for the company. What can sometimes complicate this is that for graduate roles, companies want generalists – generalists who are also specialists! Or do they want specialists who are also generalists? Be prepared for a situation in which their questions lead you to present yourself as one of the two and they then say they want the other, and even if this doesn't occur, be aware of the tension between these two requirements and use that knowledge to your advantage.

Women candidates

The system of checks and balances afforded by the behavioural interview aims to ensure that all, including female engineers, are being judged against the same set of criteria as other applicants. The engineering profession strives to increase the percentage of women that it attracts into its ranks from the current low level of about 9%. However, a recent linguistic research study of graduate engineering interviews revealed that, in common with other professions in which women are historically under-represented, there is an underlying tension between the expectation for female engineers to be both 'different' and 'the same' as males, which leads to interviewers unwittingly making it harder rather than easier for them to succeed in interviews. This phenomenon can occur even when employers are setting out to hire more women. In other words, because their expectations remain gender-circumscribed, they undermine their own pro-women recruitment agenda.[111]

In addition to filling vacancies there are two main perceived benefits of hiring more female engineers.

1. **Diversity:** trying to achieve a more balanced workforce. A core rationale for diversity is that it leads to enhanced productivity, because it brings a wider range of perspectives that in turn promote greater creativity.

2. **Communication skills:** women have a perceived superiority in communication and relational skills. These skills are taken for granted, which is not the case for men. Female engineers are implicitly associated with being good team players on account of this underlying belief in their better-developed communication skills.

Research shows that whilst men's technical skills are taken for granted, female candidates are asked disproportionately more technical questions than their male counterparts as recruiters try to match or 'fit' the candidate to their expectations of a professional engineer. Therefore, whilst a structured employment interview is designed to be objective and standardised, recruiters are still more likely to subconsciously select candidates who have things in common with them, such as outside interests and formative experiences. Consequently women need to reflect carefully about how they represent their passion for engineering through extracurricular activities and whilst the same is true for men it is less critical because of the underlying assumption of technical interest. This applies to both the interview and the CV.

What does all this mean for you? Most importantly it explains how the culture of engineering extends to recruitment situations. Probably the greatest challenge will be establishing the relevance of your experiences without feeling that you are making an unacceptable compromise in leaving certain things out (your expertise in salsa dancing, for example). Interviews have been described as 'gatekeeping' encounters – in order to get through the first gate, a little flexibility may be required. This is not so much adapting to the culture of engineering as getting yourself into a position where you can adapt the culture of engineering itself – from the inside.

Establish a relevant reason for choosing engineering

Regardless of how the interviewer phrases the question about why you want to be an engineer, this is the actual question you should be answering: 'How have your life experiences or hobbies outside of study influenced your desire to pursue a career in Engineering?'

- Have you always wanted to be an engineer? Describe how a formative experience or early role model inspired you to take up engineering.

- Did someone suggest it to you as a career? Describe what was said that initially sparked your interest and explain why. If you had never thought of engineering until your school was visited by an organisation specifically promoting engineering as an option, what ideas or ambitions did it trigger? Can you articulate exactly why you were interested in the way something worked, or was designed, or needed developing?

- Or did you decide to pursue engineering because you were good at maths? This answer lacks relevance – more relevant is that you **enjoyed** maths and, for example, can describe a time when you used it to solve a real-life problem. Most relevant would be if solving the problem also involved some kind of hands-on activity.

Demonstrate a strong orientation to engineering and create confidence in your technical and relational skills

'Desire and passion for building things' is the phrase recruiters have in mind when they consider your orientation to engineering. This helps us understand what passion for engineering actually means in practice and what you have to do (or avoid doing) to be considered passionate about it.

- Remove any distractions from your CV that might get in the way of presenting 'a coherent picture of an engineer'. Highlight extracurricular interests that help your CV look sufficiently 'engineering', if at all possible linked to what inspired you to take it up.

- When talking about your project, don't just describe it: explain what interested you and challenged you technically and why, and how you approached the challenge. This is your best chance to establish 'something in common' with your engineering recruiter.

- Be prepared for technical questions. Practise explaining concepts through the use of everyday analogies.

- Always communicate reasons for why you have done things, and where possible include your ability to 'explain why' as an integral part of your stories about teamwork, innovation, and so on; in explaining your project, in talking about your short- and long-term goals, and in asking questions about the company.

- Don't underestimate the focus on team sports as a predictor of teamwork ability. Just because women are assumed to naturally possess the so-called 'soft' skills that enable high-quality teamwork does not mean that you will not be expected to demonstrate them.

There are many local and international networks of women in engineering that can provide mentoring, motivation and support for you in your chosen career. If you're currently a student it is likely that you will be actively working to strengthen your links with each other and the senior women in your field, maintaining and building cross-disciplinary networks between the different specialisms and demonstrating through your engagement with engineering the real values of difference, diversity and equality.

It is important to raise awareness of these issues for both male and female engineers and for women to actively participate in the recruitment process in the companies where they work. There are very few women engineers in the profession who get involved in recruitment. Today, if your interviewer is a woman, she is most likely to be an HR manager sitting on a panel together with one or more male engineering managers. Remember, they are all very keen to hire you: it is up to you to make it easier for them to recognise in you a coherent picture of an engineer! Once recruited there are rewarding opportunities for women in engineering, see for example the story of Kate Cooksey in Chapter 6.

Finally there are a number of organisations that support and mentor women in engineering and related disciplines:

- Engineer Girl (www.engineergirl.org)

- Engineering Development Trust: First Edition (www.etrust.org.uk/first_edition.cfm)

- Evecracker (www.evecracker.com)

- Every Woman (www.everywoman.com)

- National Women's Network (www.national-womens-network.co.uk)

- WISE (www.wisecampaign.org.uk)

- Women in Engineering (womeninengineering.org.uk)

- Women's Engineering Society (www.wes.org.uk).

Graduate comments on the recruitment process

A group of recent graduates were asked what advice they would give based on their experience. They provided the following comments in response to the question, 'Is there any advice you would give on securing a graduate job?'

'When seeking work your passion and enthusiasm for your chosen career is paramount. Employers are looking for energy, which will come naturally where you have a real interest in a career path. Make sure you do your research about the area of work you want to enter; it is always better to be over rather than under prepared. I would also recommend being persistent, don't be afraid of chasing potential employers for opportunities – if you don't ask you don't get! Also don't forget to sell yourself and what you can bring to an employer; this can often include things you do that you may not directly relate to the attributes required of an engineer!'

Kate Cooksey (Employer: Morgan Sindall
Underground Professional Services)

'For me, choosing an industry that I am both passionate about and that is also in a key stage of development has really been a great benefit. In a recession period when many sectors were looking at cutting back and consolidating due to market constraints, I have been part of a sector that has been expanding and developing quite rapidly. In turn, this has allowed my skills to develop at the same increased rate.

'Be sure to do your research on the industry or sector you are interested in, and also the organisations you are looking at. This can help let you see what the short- to medium-term future might look like. It can really make a difference to your potential for career progression. However, do not get too caught up in long-term plans and goals, as it is important to ensure you are happy in your initial role.

Career progression is exactly that, a progression of your role, your focus and what drives you over time.

'Do not be afraid to expose yourself to the job hunting and interview processes from an early stage – it can be quite an unnerving and unnatural process for many, so getting experience can be a real benefit. Ask for feedback from interviews, whether the outcome is positive or not, as it's important to learn from your mistakes and weaknesses.'

Ed Hall (Employer: Siemens Wind Power)

'With the process of securing a graduate job, it is important to develop an interest in the sector that you want to work in and to follow it closely. Apply early to the companies that you are most interested in working for. Do your research on those companies before writing applications and before going to interviews. Selecting university projects and modules that align to the industry can also go a long way in gaining an upper hand in the job market. I would also recommend working in a wide variety of job functions early on in one's career in order to gain some perspective and to pick longer-term roles that would best suit your aspirations.'

Mukunth Kovaichelvan (Employer: Rolls-Royce)

'An engineering degree will sharpen your problem-solving and analytical abilities, both of which are important for business. However, communication, time management and leadership are also necessary to succeed. Take part in extracurricular activities and skill sessions provided by the university to develop those skills. If you're sure of opting for a business career take business courses such as marketing, finance, economics, etc. or learn through open access courses offered through websites such as Coursera and Khan Academy.

'As an engineer you will have a plethora of career options available to you, so thoroughly research the companies and roles that excite you. Actively visit career fairs to meet the firm's representatives, speak to career counsellors and make use of resources such as career guides as well as the company websites.

'No amount of theoretical knowledge can substitute for the experience of actually working for a company; so try doing internships to get a better understanding of the job. Then when you are finally applying for a full-time position you will be able to make an informed decision.'

Divya Surana (Initial Employer: Procter & Gamble)

16

Career development and progression

'For too long, engineering has been seen by others somehow as a supporting profession. I am afraid that, too often, engineers have gone along with that perception, when in fact, engineers, with their unique skillset, are leaders and need to be the drivers of change.'[1]

'But for engineers to succeed they must play more than a supporting role – they must be at the forefront of society, as leaders capable of driving change, not following it.'[1]

Responsibility for leadership in its different forms and levels develops quickly for the engineer and the role descriptors provided in this chapter are generalisations of the types that engineers may progress to.

The graduate engineer who has successfully completed their degree to master's level on an accredited programme will have satisfied the educational component towards recognition as a chartered engineer. In seeking their first appointment as a graduate engineer they will be mindful of the additional training and experience that they will require to meet the standards set by their professional engineering institution (PEI) of choice and the support provided by their new employer. Each PEI has a range of membership levels that start with student membership and which students should subscribe to for the sector insights that membership offers and because several of the schemes are free and/or waive the joining fee post-graduation. On graduation the individual progresses to the next grade of membership and works towards promotion to member, at which stage they will meet the requirements for registration as a chartered engineer. Typically this will take at least four years, although work placements during your degree may contribute to meeting the requirements. Table 5 lists the titles given to grades of membership by the PEIs discussed in this book.

Table 5: Early career PEI grades of membership

PEI	Student grade	Graduate grade
Chartered Institution of Highways and Transportation	Student member	Associate
Institute of Highway Engineers	Student member	Graduate member
Institute of Materials, Minerals and Mining	Student member	Graduate member
Institution of Chemical Engineers	Student member	Associate
Institution of Civil Engineers	Student member	Graduate member
Institution of Engineering and Technology	Student member	Associate
Institution of Environmental Sciences	Student member	Associate
Institution of Mechanical Engineers	Affiliate	Associate
Institution of Structural Engineers	Student member	Graduate member

Achieving chartered engineer status

Each PEI sets out its requirements for the professional practice and experience required to complement the graduate's educational base. Graduates will be required to demonstrate technical competence, professionalism and have exercised responsibility. Gaining chartered engineer status provides for greater professional recognition and maximises the individual's international mobility; however, there are many engineers who do not pursue chartered status and yet have worthwhile careers within the engineering sector. Notwithstanding this fact many employers expect engineers to achieve professional recognition because it is important for their clients and in some safety critical sectors, professional recognition is a requirement for progression. Most PEIs have special routes to membership and chartered status for academics based upon their research and responsibilities. At the time of writing some are actively considering routes for teaching-only staff within universities and colleges. Whilst the approach of most institutions is similar, each has its own processes outlined below and detailed on its web pages.

Chartered Institution of Highways and Transportation

Membership and chartered status is based upon an appropriate academic qualification and training and experience tested through professional review. This involves submitting a portfolio of evidence and undertaking a professional review interview.

Institute of Highway Engineers

Candidates submit details of their education, training, experience and referees' statements. This is supported by professional review and interview.

Institute of Materials, Minerals and Mining

An accredited master's degree or equivalent is combined with at least four to five years' approved training and responsible experience. Evidence is expected of the following: independent technical judgement in both practical experience and the application of scientific or engineering principles; innovation in technical matters; and direct responsibility for the management or guidance of technical staff and other resources. The submission is tested through a professional review interview.

Institution of Chemical Engineers

The Institution of Chemical Engineers grants graduates of appropriately accredited degree courses Associate Membership of the institution when they enter full-time employment or pursue a relevant postgraduate qualification. With appropriate experience graduate engineers may apply to the Institution for recognition as a chartered chemical engineer. This award is only available from the Institution and applications are peer reviewed. Candidates are expected to demonstrate depth and breadth of knowledge and to have applied these to practical applications. They will present evidence of having held responsibility for technical judgement, and having exercised best professional practice and continual professional development.

Institution of Civil Engineers

The ICE designates the Initial Professional Development as the process by which specialist skills, knowledge and competencies are achieved. These are detailed in the ICE Development Objectives. This is combined with further experience and increasing responsibility. A Chartered Professional Review considers the candidate's academic qualifications and their work experience. The review includes a written essay and interview. Successful candidates receive the protected title of chartered civil engineer.

Institution of Engineering and Technology

The IET process requires the submission of a portfolio of evidence demonstrating that the competences have been achieved and this is tested via a 15-minute presentation and one-hour interview to a panel of engineers.

Institution of Environmental Sciences

The IES is a Licensed Constituent Body of the Society for the Environment (SocEnv), enabling them to award chartered environmentalist status to those who meet the required criteria. CEnv status signifies that the holder has demonstrated 'sound knowledge, proven experience and a profound commitment to sustainable best practice within their particular profession and field of expertise.'[112] Applicants for CEnv recognition provide details of their employment history and indicate where they have demonstrated the required competencies as well as providing a report detailing relevant projects and activities that they have been involved with. This is followed by an interview.

Institution of Mechanical Engineers

Typically a graduate will need four years of experience that is recorded in the IMechE's company based Monitored Professional Development Scheme (MPDS) or

Supported Registration Scheme (SRS) where the employer does not participate in the MPDS. The MPDS provides a mentor and identifies the competences that must be demonstrated and opportunities for continuing professional development. When everything is in place the engineer submits their application for consideration and will be invited to a Professional Review Interview with a panel of engineers. The SRS scheme works in a similar way but requires the graduate engineer to identify a mentor and plan how they will demonstrate that they have achieved the required competences.

Institution of Structural Engineers

In a similar fashion to the ICE process, candidates undertake an initial professional development programme consisting of 13 core objectives. This is followed by a professional review interview and a seven-hour examination testing professional competence in structural engineering design.

Beyond chartered engineer

Having achieved full membership of a PEI and chartered engineer status, engineers, as with all professionals, are required to develop their knowledge to keep pace with current developments through continuing professional development. This may be logged and monitored by their PEI as part of their review processes. Significant achievement within the industry is recognised by most PEIs through promotion from Member to Fellow of the institution.

Career structure

The role of an engineer in industry can take many forms and may, for example, focus on a specialist technical field, project management, business or management functions. It is therefore difficult to generalise about a typical career structure, beyond the expectation that progression will normally see an increase in responsibility for technical detail, health and safety, supervision of staff and/or budgets.

Engineer

Graduate engineers will complete their initial training and then take responsibility, under experienced supervision, for activities of greater complexity and value. They will have responsibility for delivering these within budget and to deadlines whilst achieving the required specification and quality. The engineer will use their knowledge and skills of design and analysis in the development, implementation and operation stages of the project. Roles may involve contributing to the development of contract specifications, pricing, etc. Good employability skills are needed for effective team-working and communication with interested parties. The range of activities is diverse and will depend upon the particular industrial sector and role in which the engineer is employed.

Senior engineer

With experience and an ability to synthesise new ideas and techniques to achieve a project, the engineer may move into a senior engineer role. Their experience and ability will be proven and they will supervise others whilst drawing upon the technical expertise of colleagues as required. Responsibility will extend to larger projects, planning and budgets and they will have an overview of the project. They will need excellent communication skills and may produce technical position papers to promote their company's operations or products.

Principal engineer

Taking greater responsibility for specifying complex projects and their requirements, the principal engineer may then divide these into manageable sub-projects or work packages for the engineering team. The principal engineer's experience and expertise equips them to manage uncertainty and innovation in the design or project and to contribute to the development of business plans.

Engineering director/chief engineer

This position describes those with a proven engineering expertise and an ability to innovate. The engineering director or chief engineer will have some responsibility for the strategic direction of the engineering activity and may be concerned with the adoption of emerging technologies or new materials to improve existing operations or products and to establish new ones. This requires a broad knowledge and understanding of the sector and the company's main competitors and an ability to assimilate and interpret market information.

Technical manager

Engineers who choose to follow a management career will use their skills to organise and direct people, business processes, tasks and other resources. They will inform their management skills and execution with their technical understanding of the project or business.

Project manager

The project manager uses their experience of project management to assess the business case for new projects. They will have an understanding of the strategic direction and business needs of the company and will advise on the risks to the company associated with the proposed project and their mitigation.

Programme director

The programme director role is the management equivalent of the engineering director, taking a strategic responsibility for developing or introducing new programme management techniques and monitoring the sector for opportunities and competitor performance.

My story . . .

Divya Surana, Business Analyst at Procter & Gamble and Brand Manager at Fizzy Foodlabs

My first role in P&G was that of a Site Services Leader (SSL) of its UK headquarters. I was responsible for managing service providers such as British Telecom (telephony), Xerox (printing), Hewlett Packard (computers and IT services) and Jones Lang LaSalle (real estate, catering and security). Together we delivered projects that were crucial to the running of the business.

After a year, I was promoted to SSL, North UK – covering three offices with 1,500 employees. I was also entrusted with delivering initiatives across

Western Europe, which I passionately led, involving people from different functions, companies, geographies and hierarchies. My next role was that of a Global Business Analyst where I led a project to make P&G's tax reporting practices more efficient in over 60 countries.

At P&G the high level of responsibility bestowed on me from day one was incredible. Being able to manage external partners who were double my age and had 20 years of work experience was quite a confidence boost. I was also fortunate enough to work directly with senior management (including the chief executive officer and chief finance officer) and those interactions were extremely insightful.

The opportunity to travel and meet with colleagues from all over the world was a great perk. I've been on business trips to Hong Kong, Philippines, India, Switzerland, Italy, Germany, the USA and Costa Rica.

After three years at P&G, I decided to take the plunge into entrepreneurship and move back home to India. I'm currently a Brand Manager at Fizzy Foodlabs, which is a start-up company that's looking to launch ready-to-eat and ready-to-cook food products in India. Building a brand identity, promoting the products via websites, social media and advertising as well as selling the products online and in-store are my key priorities.

At Fizzy Foodlabs I get to mix my passion and talent, plus there is always something new to discover. Although my core function is marketing, I'm involved in sales, business development and human resources. As an entrepreneur you are essentially starting and running a business from scratch so it's a high-pressure job, but equally rewarding. The prospect of my company becoming the next Innocent Smoothies or Gü Puds is truly exciting.

17

Postgraduate study

'Comparison of these doctoral graduates' earnings six months
after graduation and three years later indicate that doctoral
graduates continued to enjoy a salary premium relative to
masters and good first degree graduates. . . . Doctoral
graduates consistently perceive positive impact of their
doctoral degree experience on their workplace activities, career
progression and to a lesser extent their wider lives. Within
the workplace this experience helps them to be more innovative
and influence the work of others, as might be expected of
those with very high-level knowledge. Many also report that
it enables them to change organisational culture and working
practices, presumably reflecting their high competency levels.'[113]

Postgraduate study refers to degrees that are at level 7 or higher in the National
Qualification Framework (NQF) or level 11 in Scotland, as described in Chapter 3.
For those wishing to follow a career in engineering the educational requirement is
for a degree at level 7 that has been accredited by the appropriate PEI. For many
this is achieved through an integrated Master of Engineering degree. For those

graduating with a bachelor's degree there is a need to make up the level 7 material and one means is to study for a taught MSc degree in an appropriate subject. For those choosing a non-engineering career they may find that they need to undertake a higher degree in their chosen area. Some career paths, such as those within higher education, may require a research degree. Not all career routes require a postgraduate qualification, but having one will increase the opportunities for recruitment, especially in competitive sectors.

There are also widespread opportunities for postgraduate study internationally. There are, for example, several hundred engineering MSc degrees available across mainland Europe that are taught in English and have very competitive fees in comparison to the UK. Likewise, opportunities for research degrees exist in many other countries and may be attractive to those who wish to use the opportunity to study in another culture or laboratory recognised internationally for the excellence of its research.

Types of postgraduate study in engineering

Postgraduate study may be pursued via either taught or research degrees.

Taught degrees

Integrated Master of Engineering (MEng) degrees are fixed at level 7 in the NQF (level 11 in Scotland), but are rarely thought of as a taught postgraduate degree because they are integrated with and extend the level 4–6 material associated with the Bachelor of Engineering degrees. They also benefit from an undergraduate funding regime. Frequently a MEng degree course will have a broader curriculum than that offered for an MSc.

Master of Science (MSc) degrees normally last for 12 months and tend to follow a focused curriculum on a specialist area of engineering. Some may be studied on a part-time basis over two or three years. They do not require that a student progress immediately from their bachelor's to the master's degree, providing an opportunity for some students to gain industrial experience before returning to their studies. At the time of writing the funding regime is less favourable than that for the integrated master's degree and you should be prepared to pay full course fees and cover your living costs. In some sectors experiencing a skills shortage, such as aerospace engineering or tunnelling, scholarships may be available from industry and/or government. The MSc degrees are better understood internationally in comparison to the MEng degree. For students who do not meet the entry requirements for a taught MSc, some universities offer a graduate diploma that aims to bridge the gap.

An MSc requires 180 credits to be successfully completed for the award of the degree. For those not wishing to undertake the 180 credits, some courses offer the lesser awards of **Postgraduate Certificate** (90 credits) or **Postgraduate Diploma** (120 credits).

Research degrees

Admission to a research degree normally requires a minimum of a 2.i classification in the first degree or a 2.ii classification supplemented by an MSc. The prospective university and supervisor will look for ability in research evidenced by individual project work within the first degree and MSc. The choice of where to apply to will be governed by the subject to be studied and the universities hosting good research groups working in that area. This can be ascertained by looking at university websites, research journals and the research funding agencies' lists of awards. It is important to choose both a research group that has a good standing in the research community and a supervisor with whom you can establish a good working relationship. If feasible try to meet your prospective supervisor and talk to some of their research students about their experience.

Master of Research (MRes) degrees are of 12 months' duration and combine subject teaching with training in research techniques, including a substantial research project. They are normally designed as a route to a doctorate degree.

Master of Science by Research (MSc) and Master of Philosophy (MPhil) degrees involve two years of in-depth research in an area of interest, resulting in the submission of a thesis. They involve some training in research methods.

Doctor of Philosophy (PhD) degrees consist of normally three to four years full-time (or equivalent) research into an area of specialist interest. The award is based upon the final written thesis tested via a viva (an oral exam). In engineering disciplines the opportunities for PhD study are often linked with research programmes funded by industry or other agency and therefore some student funding may be available within the award.

NewRoutePhD™ is an 'integrated PhD' designed to underpin research with a structured programme of formal training, including subject specific modules, research skills training and generic employability skills. The added content extends the period of study to four years.

Engineering Doctorate (EngD) degrees combine academic research in an industrial setting with taught modules in related subjects. Whereas a PhD is the traditional route to an academic career, amongst others, the EngD is designed for those hoping to work in industry or other non-academic situations.

In each of the three doctorate degrees above the test of success is original research leading to new knowledge.

My story . . .
Navroop Singh Matharu, research student at University of Warwick

I graduated with a MEng (Hons) in Civil Engineering in 2010. As an undergraduate I had undertaken a research placement through a university research internship which funded 12 weeks over one summer. I worked with a member of staff on computational fluid dynamic simulations of auger piles penetrating deep clay. Through this I engaged with the research world, developing analytical skills, evaluating data sets and independent thinking. I really enjoyed having the time to examine an individual problem in depth and to explore the increasing numbers of questions that arose from the process. I also saw the way that research could lead to new understanding that had practical application in industry.

Upon graduation I was awarded a PhD studentship to research connections and joints in fibre reinforced polymer (FRP) buildings and bridges. This builds on the core knowledge and skills from my first degree and allows me to develop an in-depth expertise in the use of a novel material for construction.

What excites me about research is the fact that what I am doing is new; nobody else has done this and the results will feed through to construction design codes in the USA and possibly Europe. The discipline of research forces you to stretch your mental ability to understand the problem and develop appropriate methodologies to test your theories. A side benefit has been the opportunity to travel to places like Rome, Belfast and Vancouver to meet other experts in the field and share ideas. I also enjoy the physical testing of structures in the research laboratory and having a good degree of autonomy in managing my research programme and working day.

Longer term I hope to find a role within a research environment in an industrial setting and to exploit the growing trend of using FRP in construction.

If asked to give advice to somebody considering a research degree I would say that you need to be confident that this is what you want, because you will commit three years of your life to it. You should carefully choose your supervisor to be somebody you can have a good rapport with and find a topic that interests you. Research is not all plain sailing – it has its ups and downs and in the low moments it is important to have that underlying resolve to achieve your goal. I have found that good time management is the key to success in research, especially when performing laboratory experiments. When research is successful in developing new knowledge the feeling of satisfaction is fantastic!

18

Alternative careers with a degree in engineering

> 'There is good econometric evidence that the demand for graduate engineers exceeds supply and the demand is pervasive across all sectors of the economy. The implication of this is that the economy needs more graduate engineers for both engineering and non-engineering jobs. The evidence can be seen in a persistent, sizeable wage premium for people holding engineering degrees and this premium has grown over the last 20 years.' [114]

Engineering degrees are designed to meet a vocational need, but, as with all degree courses, they provide an education and employability skills that are desirable to many employers, both internal and external to the engineering sector. Sometimes the choice of graduate career is impacted by external factors, such as an economic downturn, and graduates may need to diversify their search criteria for a graduate position.

According to the CBI[4] 41% of employers favour graduates from science, technology, engineering or mathematics (STEM) degrees. This highlights the broad range of employment opportunities that engineers enjoy beyond the diversity of engineering roles. Engineers have a broad skillset that transfers well to other disciplines where systematic analysis of a problem or an understanding of technical vocabulary is required. Examples of alternative careers are described below, but this is not a comprehensive list.

Accountant

Engineering graduates are attractive to many accountancy firms not just because of their skills in numeracy, analysis and the use of IT, but also because many accountants deal with industrial concerns and therefore an understanding of the technical operations and products can be very helpful in performing their duties. Accountants work in all sectors in the following roles.

- Financial accountants managing day-to-day financial transactions within the company or organisation, performing internal audits, managing investments, protecting assets and advising the directors.

- Management accountants analysing and modelling the financial data of the organisation. In some situations their engineering background can help place the assessment of future projects into context, in for example cost–benefit analyses.

- Auditors who examine the practices and accounts of companies and organisations to ensure that they follow best practice and are both robust and an honest reflection of the reality.

- Specialist advisers in areas such as taxation, corporate finance and insolvency.

Actuaries and insurance professionals

These use their analytical skills to assess the probability of something happening (risk) and use that data to set, for example, the cost of insurance or pension contribution. Many large engineering projects, such as tunnels, or products, such as ships, will carry insurance against failure or loss and therefore an engineering background can assist in understanding the nature of the risks associated with the project or product.

Careers adviser

Careers advisers provide assistance to people as they identify their future education, training and career. This will involve working with both individuals seeking information and employers or recruiters seeking personnel. For example all universities employ career advisers to: support students and graduates in finding employment or further education; liaise with recruiters to promote their opportunities to the target audience; and to advise academic staff about any issues affecting the employability of their students.

Civil service

With over 250 departments and non-departmental public bodies (NDPB) the civil service employs 444,000 people across the UK.[115] There is a broad range of opportunities for new graduates and experienced professionals alike, including roles in departments with readily apparent engineering interests such as the Defence Engineering and Science Group, Environment Agency, Health and Safety Executive and Highways Agency. The role of the civil servant is to provide support for the government of the day through, for example, research, report writing, public engagement activities and delivering government policy.

Entrepreneur

In 2009/10 4.8% of all graduates operated on a freelance basis or were self-employed, and this group was steadily growing in size.[49] Increasingly, university engineering students have access to modules dealing with entrepreneurship and support structures that can provide mentoring and support in attracting funding.

Investment banker

Investment bankers provide advice to client firms on the acquisition of new assets, floating on the stock market and dealing in shares. Investment banks also deal in the normal banking transactions for companies, international corporations and governments. Investment banking fields may specialise in areas such as technology where a background in engineering may be useful.

Management consultant

Management consultants provide advisory services to client firms such as product line management and strategy. Management consulting is specialised by industry, offering engineering graduates an opportunity to use their knowledge. Management consultants will be contracted by companies to address a particular problem, perform a feasibility study or improve an existing process. The engineer's ability to analyse a process or product, disassemble it and create an improved version transfers easily to the role of management consultant.

Officer in the armed forces

There are a wide range of opportunities for those willing to enter the armed forces, ranging from roles on the front line of combat, peace keeping or humanitarian relief activities to engineering support roles, logistics and other support services.

Teaching

Engineering graduates and experienced engineers have much to offer the teaching profession. Engineering subjects are taught within some schools, but all will offer subjects familiar to the engineering graduate: mathematics, physics and design and technology. There are national shortages in some of these subject areas, especially physics. In response the Institute of Physics and Royal Academy of Engineering have jointly supported the idea of meeting the shortfall by encouraging engineering students to become teachers of physics and the IOP offers scholarships worth £20,000.[116] One popular route into teaching for graduates is via Teach First, which offers leadership training and a route to a professional teaching qualification over two years.

Technical author

There are a wide variety of outputs for technical authors, including writing for popular or specialist journals or magazines, authoring technical manuals or online help files, books or help desk procedures. Technical writers combine excellent writing skills with an ability to assimilate new knowledge, often working to a deadline. They will need good interpersonal skills for interviewing others and knowledge of appropriate industry-standard software and techniques.

University administrator

Universities are large and complex businesses that are supported by administrators who work either in the central functions or within individual departments. Administrative staff will manage teams responsible for functions such as student admissions, accommodation, estates management, international activities, catering, welfare and management of human resources. They will often have several roles, including acting as secretaries to various academic committees. Those located within academic departments may be responsible for supporting the taught or research degrees, finance and bids for research funding and the resulting awards. They will work closely with their academic colleagues. Good team-working, communication, IT, time management and commercial skills are required.

Concluding remarks

When applying to any of the aforementioned fields, one should stress technical and analytical abilities.

My story . . .

Gaston Chee, self-employed, co-founder of BeGo (an internet start-up company funded by venture capitalists)

My co-founder, Jennifer Leong, and I travelled from Malaysia to study at university in the UK. Our separate experiences were similar. We found ourselves in a completely foreign environment with very limited luggage allowance and no friends and family around us. The shops and surroundings were all unfamiliar to us; we had no SIM card with which to call home; we didn't have anything to sleep under or anything to cook with. Then we were faced with the task of setting up a bank account, phone contract and insurance policy – some of our friends were unlucky as they had their things stolen on their way to the UK. To make things harder, we were not used to speaking English; we couldn't understand the accent and didn't know the technical terminology involved in setting up UK accounts and policies. The idea for BeGo grew out of our dissatisfaction with our experience. We are dedicated to serving international students by making their transition to the UK as easy as possible, by providing services such as insurance and start-up packs. BeGo enables them to enjoy the best the UK has to offer.

Whether you are aiming to become a lawyer, doctor, engineer or a business owner, you just have to bear one thing in mind – there are millions of people who can do your job. Landing a great career is not about having a great CV and graduating with a First Class honours degree, in the same way that a camera alone does not make a great picture or a typewriter write a great novel. The bottom line is that a career is about you and not about the career.

The modern marketplace demands that people possess a wide range of skills. But what core qualities are truly essential to career advancement, regardless of industry or job? The fact is there is no shortage of people with skills. The key is to have the right attitude.

I started my company in a foreign country when I was 24, raised nearly a million dollars from strangers and here is what life has taught me:

Learning: Prioritise learning and develop a spirit that continuously learns. Think of yourself as a work in-progress and invest in yourself every single day and the learning will never stop. Edmund Hillary made several unsuccessful attempts at scaling Mount Everest before he finally succeeded. After one attempt he stood at the base of the giant mountain and shook his fist at it. 'I'll defeat you yet,' he said in defiance, 'because you're only as big as you're going to get – but I'm still growing.' My initial attempts to raise funds for my business idea repeatedly failed, but every time I failed, I learned. And every time I learned, I grew and I tried again until one day the deal was sealed.

Be real and be passionate: During my final year at university, many of my friends were desperate to join either one of the big four accountancy firms or an investment bank. However, when asked what their main motivation was, their simple response was: 'That's what everyone else is doing' or 'That's where they pay the highest graduate salary and if you're lucky they even pay for your master's!' Going into a marketplace without knowing your likes, strengths and passions in my opinion is not going to succeed. In the long run, it will not be sustainable and every so often, you will break down.

In her commencement speech at Harvard University, J.K. Rowling said these words: 'So given a Time Turner, I would tell my 21-year-old self that personal happiness lies in knowing that life is not a check-list of acquisition or achievement. Your qualifications, your CV, are not your life, though you will meet many people of my age and older who confuse the two. Life is difficult, and complicated, and beyond anyone's total control, and the humility to know that will enable you to survive its vicissitudes.'

Some people will tell you to find your passion and then pursue it wholeheartedly. Such advice has serious merits but also huge drawbacks.

One of the main drawbacks is that they presume a static world and associate passion with some sort of magical ingredient that will eventually lead you to success, but as we all know you change, the market changes, the competition changes and the world changes!

A good way to tell if you have a passion for what you are working on is to ask this: Do your weekends look a lot like your weekdays? When you get home from work, do you find yourself wanting more? Do your ears perk up whenever someone talks about a certain subject? Do you feel compelled to ask questions and really listen to the answers? That's passion, and it fuels a competitive advantage that cannot be faked. It is what drives you to work when you do not need to or think about new solutions to old problems when everyone else is spending the weekend flipping the 'work' switches off.

Integrity: Most recently, my company BeGo has started recruiting. I was reminded again of what Warren Buffett (Chairman and CEO of Berkshire Hathaway) has to say about hiring: 'In looking for people to hire, look for three qualities: integrity, intelligence, and energy. And if they don't have the first one, the other two will kill you.' Over the past two years of running BeGo, I have had the opportunity to meet many different people and I have found that integrity in supposedly intelligent people is one of the rarest treasures that you will ever find. The value of trust others have in you is far beyond anything that can be measured. As an employee, it means your boss is willing to trust you with additional responsibilities. For entrepreneurs, it means investors are willing to trust you with their hard earned money.

Glossary

3D printing	A process by which a solid object is created by adding multiple layers based on a computer model. In so doing it differs from traditional machining techniques that generally remove material to create the shape
Aerodynamic drag	A force generated by the interaction between a solid and a fluid with which it has contact, as one moves relative to the other
AFEO	ASEAN Federation of Engineering Organisations
Agrochemicals	Chemicals used in agriculture
APEC	Asia Pacific Economic Cooperation
Aquifer	An underground layer or structure of porous and permeable rock or sediment that holds water and from which the water can be extracted
ASEAN	Association of Southeast Asian Nations
BEng	Bachelor of Engineering degree
British Standards	Documents published by the British Standards Institution that set out agreed procedures for activities or agreed definitions
Brownfield site	Land, typically urban, identified for redevelopment that may have a history of use which has contaminated the ground. Examples include industrial plants, gas works and waste tips
BSc	Bachelor of Science degree
CBI	Confederation of British Industry
CEC	Commonwealth Engineers' Council
CEng	Chartered Engineer
Chromatography	A range of techniques to separate compounds by allowing a solution or mixture to seep through an adsorbent medium such that each compound becomes adsorbed into a separate, often coloured, layer

Glossary

CIHT	The Chartered Institution of Highways and Transportation
CPD	Continuing personal and/or professional development
Distillation	A process that uses the variation in the vaporisation point of substances to separate them from each other
Dye pen testing	A non-destructive test for surface flaws and cracks in components involving a dye which is applied to the surface. The dye is drawn into any cracks or flaws by capillary action, the surface is wiped clean and an agent applied that draws any dye back to the surface, revealing the crack or flaw
ECUK	The Engineering Council UK
ENAEE	European Network for Accreditation of Engineering Education
Engineering drawing	A detailed diagram of a component or assembly, drawn to scale
EngTech	Engineering Technician
ERASMUS	European Community Action Scheme for the Mobility of University Students is a European Union student exchange scheme
EUR-ACE	European Accredited Engineer
Fail-safe	Refers to a device or system that ensures that when something fails it does so in a manner that prevents the failure causing harm or damage
FEANI	European Federation of National Engineering Associations
Feedstock	Any renewable, biological material that can be used directly as a fuel, or converted to another form of fuel or energy product
FEISEAP	Federation of Engineering Institutions of Southeast Asia and the Pacific
Fly ash	Fine by-product of combustion, for example from coal fired power stations
FTSE	An independent company owned by London Stock Exchange Group
Gross value added (GVA)	An economic measure of the value of goods and services produced in an area, industry or sector of an economy
Ground investigation	Similar in nature and purpose to a site investigation but limited in scope to the ground conditions and properties

HNC	Higher National Certificate
HND	Higher National Diploma
ICE	The Institution of Civil Engineers
IChemE	Institution of Chemical Engineers
ICT	Information and communications technology
ICTTech	ICT Technician
IEng	Incorporated Engineer
IET	The Institution of Engineering and Technology
IHE	The Institute of Highway Engineers
IMechE	The Institution of Mechanical Engineers
IOM3	Institute of Materials, Minerals and Mining
IStructE	The Institution of Structural Engineers
MEng	Master of Engineering degree, described as an integrated master's because they are closely linked to the BEng degrees
MPDS	Monitored Professional Development Scheme
MSc	Master of Science degree
Nanomaterial	A natural, incidental or manufactured material containing particles, in an unbound state or as an aggregate or as an agglomerate and where, for 50% or more of the particles in the number size distribution, one or more external dimensions is in the size range 1 nm–100 nm. In specific cases and where warranted by concerns for the environment, health, safety or competitiveness the number size distribution threshold of 50% may be replaced by a threshold between 1% and 50%[117]
Nanotube	Hollow cylindrical molecule usually composed of carbon atoms
NQF	National Qualifications Framework for the UK excluding Scotland
NVQ	National Vocational Qualification
PEI	Professional Engineering Institution
Phase 1 Desk Studies	The first phase of a site investigation comprising a desk study in which existing site information and data are compiled from repositories such as the British Geological Survey
Photonics	The science of the harnessing of light. Photonics encompasses the generation of light, the detection

	of light, the management of light through guidance, manipulation and amplification, and most importantly, its utilisation for the benefit of mankind (Pierre Aigrain, 1967)
Plastic electronics	Circuits printed onto almost any surface, including flexible sheets
Power electronics	The application of electronics to the efficient management of power from milliwatts to gigawatts
Powertrain	A system that generates and transmits power, for example in a car this includes power generation in the engine, through the transmission down to the driving contact with the road surface
PSRB	Professional, statutory and regulatory body
Radiation hardened	Electronic components and systems that are resistant to damage and malfunction as a result of ionising radiation
RAEng	Royal Academy of Engineering
Rapid prototyping	Techniques used to quickly fabricate an item from a computer-based model
Reaction	As in a chemical reaction: a process that transforms one substance into another
Safety critical	A device or system that if it malfunctioned would endanger equipment or people. Examples include computer control systems for aircraft or nuclear power stations
SCQF	Scottish Credit and Qualifications Framework
Semta	The Sector Skills Council for Advanced Manufacturing and Engineering
Sensor	A device that measures a physical quantity and generates a signal that may be used for monitoring behaviour
Site investigation	The investigation of sites for the purposes of assessing their suitability for the construction of civil engineering and building works and of acquiring knowledge of the characteristics of a site that affect the design and construction of such work and the security of neighbouring land and property (BS5930 1999)
Small-cap	Small-cap stock refers to a company with a market capitalisation (calculated by taking a firm's current share price and multiplying that figure by the total number

	of shares outstanding) near the low end of the publicly traded spectrum[118]
STEM	Collective abbreviation for science, technology, engineering and mathematics
SVQ	Scottish Vocational Qualification
Trial pits	Trenches excavated for the purpose of ground investigation
Twin-bore tunnels	A pair of tunnels excavated in parallel, for example to isolate trains or cars travelling in opposite directions
UK-SPEC	UK Standard for Professional Engineering Competence
UNESCO	United Nations Educational, Scientific and Cultural Organisation
UNIDO	United Nations Industrial Development Organization
Unit operation	A single step in a process that results in a physical change to the target of the process, for example freezing water to ice is a unit operation
WFEO	World Federation of Engineering Organisations

Endnotes

1 Browne (2011). *For the Engineering Leaders of Tomorrow: Two Lectures by Lord Browne of Madingley. President, The Royal Academy of Engineering 2006–2011.* London: The Royal Academy of Engineering.

2 National HE STEM Project (2012). *Unemployment of Engineering Graduates: The Key Issues.* Birmingham: National HE STEM Programme. Retrieved 3 April 2013 from www.epc.ac.uk/wp-content/uploads/2012/11/HE-STEM-Engineering-Grad-Unemployment-2012.pdf

3 EngineeringUK (2013). *The State of Engineering.* London: EngineeringUK. Retrieved 12 December 2012 from www.engineeringuk.com/Research/Engineering_UK_Report/

4 CBI (2011). *Building for Growth: Business Priorities for Education and Skills: Education and Skills Survey 2011.* London: CBI. Retrieved 25 January 2012 from www.cbi.org.uk/media/1051530/cbi__edi_education___skills_survey_2011.pdf

5 Rose, J. (2007). *Why Manufacturing Matters.* The Gabor Lecture 2007, Imperial College. Retrieved March 7, 2013 from www3.imperial.ac.uk/events/dennisgaborlecture

6 Bailey, N., and Rodriguez-Vega, D. (eds) (2010). *Engineering a Better World: Conclusions from the CEC/ICE 2010 Commonwealth Week Conference.* London: Commonwealth Engineers' Council and The Institution of Civil Engineers. Retrieved 6 December 2012 from http://cec.ice.org.uk/documents/Report_CEC_Version_(2).pdf

7 UNISDR (2011). *Proceedings of the Third Session of the Global Platform for Disaster Risk Reduction and World Reconstruction Conference.* Geneva, Switzerland, 8–13 May 2011, New York: United Nations. Retrieved 3 April 2013 from www.preventionweb.net/go/22420

8 Glenn, J.C., Gordon, T.J., and Florescu, E. (2012). *2012 State of the Future: Executive Summary.* The Millennium Project. Retrieved 3 April 2013 from www.millennium-project.org/millennium/2012SOF.html

9 WWAP (World Water Assessment Programme) (2012). *The United Nations World Water Development Report 4: Managing Water under Uncertainty and Risk.* Paris: UNESCO. Retrieved 3 April 2013 from www.unesco.org/new/en/natural-sciences/environment/water/wwap/wwdr/wwdr4-2012/

10 Brown, O. (2008). *Migration and Climate Change.* Geneva: International Organization for Migration. Retrieved 3 April 2013 from www.publications.iom.int/bookstore/free/MRS-31_EN.pdf

Endnotes

11 Lohmann, J., Rollins Jr, H., and Hoey, J. (2006). Defining, developing and assessing global competence in engineers, *European Journal of Engineering Education*, 31(1), pp. 119–31, Education Research Complete, EBSCOhost, Retrieved 14 July 2013.

12 www.etrust.org.uk/headstart.cfm

13 www.etrust.org.uk

14 ENAEE (2011). *Statutes*. Brussels: European Network for Accreditation of Engineering Education. Retrieved 15 December 2012 from www.enaee.eu/wp-content/uploads/2012/01/ENAEE-statutes-revised-Oc-20111.pdf

15 www.ice.org.uk/About-ICE/What-we-do/Commonwealth-Engineers-Council

16 www.tomorrowsengineers.org.uk/careers/routes.cfm

17 www2.ofqual.gov.uk/files/2011-08-22-qualifications-leaflet-rough-guide.pdf

18 www.ocr.org.uk/qualifications/other-general-qualifications-mathematics-for-engineering-level-3-certificate-h860

19 www.theengineer.co.uk/channels/skills-and-careers/wmg-supports-degree-provision-for-jaguar-land-rover-staff/1015236.article

20 IET (2012). *Furthering Your Career – A Report from the Institution of Engineering and Technology*. London: The Institution of Engineering and Technology. Retrieved 4 April 2013 from www.theiet.org/business/accreditation/downloads/career-report-synopsis.cfm

21 ECUK (2013). *UK Standard for Professional Engineering Competence: Engineering Technician, Incorporated Engineer and Chartered Engineer Standard*. London: Engineering Council UK. Retrieved 4 April 2013 from www.engc.org.uk/education--skills/engineering-gateways

22 RAEng (2007). *Educating Engineers for the 21st Century*. London: The Royal Academy of Engineers.

23 Innovation, Universities, Science and Skills Committee (2008). *Minutes of Evidence: Engineering: Turning Ideas into Reality: Q132*. London: House of Commons. Retrieved 4 April 2013 from www.publications.parliament.uk/pa/cm200809/cmselect/cmdius/50/8043008.htm

24 Finniston, M. (1980). *Engineering our Future*. Report of the Committee of Enquiry into the Engineering Profession, Sir Montague Finniston (Chair). London: HMSO.

25 CBI (2012). *Learning to Grow: What Employers Need from Education and Skills – Education and skills survey 2012*. London: CBI. Retrieved 3 April 2013 from www.cbi.org.uk/media/1514978/cbi_education_and_skills_survey_2012.pdf

26 Mughal, H. (2004). Personal Communication.

27 Wilson, T. (2012). *A Review of Business–University Collaboration*. London: Department for Business Innovation and Skills. Retrieved 28 February 2012 from www.gov.uk/government/uploads/system/uploads/attachment_data/file/32383/12-610-wilson-review-business-university-collaboration.pdf

28 www.imeche.org/about-us/our-vision. Retrieved 25 February 2013.

29 www.molinsitcm.com/innovative-packaging/pyramid-tea-bags. Retrieved 25 February 2013.

30 SMMT (2013). *Motor industry facts 2013*. London: The Society of Motor Manufacturers and Traders Limited. Retrieved 26 February 2013 from www.smmt.co.uk/wp-content/uploads/SMMT-2013-Motor-Industry-Facts-guide.pdf?9b6f83

31 National Careers Service (2012). *Job Profiles: Automotive Engineer*. Retrieved 26 February 2013 from https://nationalcareersservice.direct.gov.uk/advice/planning/jobprofiles/Pages/automotiveengineer.aspx

32 www.building.co.uk/skills-shortage-deepens-services-hit-hard/3103785.article. Retrieved 26 February 2013.

33 ICE (2009). *Terms of Reference for Institution and Inter-Institution Medals*. London: Institution of Civil Engineers.

34 IStructE (2013). *What Do Structural Engineers Do?* London: The Institution of Structural Engineers. Retrieved 7 January 2013 from www.istructe.org/education/structural-engineering-explained

35 www.theihe.org. Retrieved 7 January 2013.

36 www.thamestunnelconsultation.co.uk

37 http://data.nce.co.uk/consultants/Default.aspx. Retrieved 7 January 2013.

38 IEEE (2001). *Your Career in the Electrical, Electronics, and Computer Engineering Fields*. Institute of Electrical and Electronics Engineers, Inc. Retrieved 10 July 2013 from www.ieeeusa.org/careers/yourcareer.html

39 BIS (2012). *Economics Paper No. 18: Industrial Strategy: UK Sector Analysis*. London: Department for Business Innovation and Skills. Retrieved 1 April 2013 from www.bis.gov.uk/assets/BISCore/economics-and-statistics/docs/1/12-1140-industrial-strategy-uk-sector-analysis.pdf

40 BIS (2011). *Power Electronics: A Strategy for Success*. London: Department for Business Innovation and Skills. Retrieved 1 April 2013 from www.gov.uk/government/uploads/system/uploads/attachment_data/file/31795/11-1073-power-electronics-strategy-for-success.pdf

41 TSB (2013). *Electronics, Sensors and Photonics*. London: Technology Strategy Board. Retrieved 1 April 2013 from www.innovateuk.org/ourstrategy/our-focus-areas/electronicsphotonicsandelectricalsystems.ashx

42 Engineering the Future (2011). *Infrastructure, Engineering and Climate Change Adaptation: Ensuring Services in an Uncertain Future*. London: The Royal Academy of Engineers. Retrieved 20 November 2012 from www.raeng.org.uk/news/publications/list/reports/Engineering_the_future_2011.pdf

Endnotes

43 Campbell, J., Goldstein, S., and Mowry, T. (2006). *Cyber-Physical Systems*. Position Paper. National Science Foundation Workshop on Cyber-Physical Systems held on October 16–17 2006 in Austin, Texas. Retrieved 1 April 2013 from http://varma.ece.cmu.edu/cps/Position-Papers/Goldstein-Mowry-Campbell.pdf

44 IBM (2010). *IBM Global Business Services: Telecommunications*. Retrieved 1 April 2013 from http://public.dhe.ibm.com/common/ssi/ecm/en/gbe03259usen/GBE03259USEN.PDF

45 PELG (2012). *UK Plastic Electronics Capability Guide*. Electronics, Sensors, Photonics Knowledge Transfer Network in collaboration with the Plastic Electronics Leadership Group. Retrieved 1 April 2013 from http://ukplasticelectronics.com/wp-content/uploads/2012/09/PE_CapabilityGuide_V1prJun12.pdf

46 www.pragmaticprinting.com

47 www.plusplasticelectronics.com/RetailPackaging.aspx

48 www.plusplasticelectronics.com/Buildingsinfrastructure.aspx

49 HECSU/AGCAS (2013). *What Do Graduates Do? 2013*. Higher Education Careers Service Unit. Retrieved 18 March 2013 from www.hecsu.ac.uk/what_do_graduates_do_archive.htm

50 www.nationalgridcareers.com/Our-Business-Areas/Transmission/Assests-Management

51 www.prospects.ac.uk/electronics_engineer_salary.htm. Retrieved 18 March 2013.

52 IChemE (2013). *Why not Chem Eng? What Is Chemical Engineering?* Institution of Chemical Engineers. Retrieved 2 March 2013 from www.whynotchemeng.com/information/what%20is%20chemical%20engineering.aspx#.Ufltg42G2Vw

53 IEA (2012). *World Energy Outlook 2012*. Paris: International Energy Agency.

54 BP plc (2013). *BP Energy Outlook 2030*. London: BP plc. Retrieved 1 March 2013 from www.bp.com/content/dam/bp/pdf/statistical-review/BP_World_Energy_Outlook_booklet_2013.pdf

55 IChemE (2012). *Salary Survey 2012*. Institution of Chemical Engineers. Retrieved 1 March 2013 from www.icheme.org/~/media/documents/icheme/media%20centre/summarypage.pdf

56 AIAA (2009). *What is Aerospace Engineering?* American Institute of Aeronautics and Astronautics, Inc. Retrieved 18 March 2013 from https://info.aiaa.org/AskPolaris/aero/Pages/default.aspx

57 BIS (2013). *Lifting Off: Implementing the Strategic Vision for UK Aerospace*. London: Department for Business Innovation and Skills. Retrieved 18 March 2013 from www.gov.uk/government/uploads/system/uploads/attachment_data/file/142625/Lifting_off_implementing_the_strategic_vision_for_UK_aerospace.pdf

58 CBI (2012). *Playing our Strongest Hand: Maximising the UK's Industrial Opportunities*. London: CBI. Retrieved 3 April 2013 from www.cbi.org.uk/media/1821466/cbi_industrial_strategy_report.pdf

59 Witts, S. (2013) *Towards a UK Aviation Skills Plan*. London: Royal Aeronautical Society. Retrieved 3 April 2013 from http://aerosociety.com/Assets/Docs/Publications/ DiscussionPapers/Towards_A_UK_Aviation_Skills_Plan.pdf

60 A|D|S (2011). *UK Aerospace Industry Survey 2011*. A|D|S. Retrieved 3 April 2013 from www. adsgroup.org.uk/community/dms/download.asp?txtPageLinkDocPK=34883

61 A|D|S (2012). *UK Aerospace Industry Survey 2012*. A|D|S. Retrieved 3 April 2013 from www. adsgroup.org.uk/community/dms/download.asp?txtPageLinkDocPK=34884

62 PwC (2012). *Aerospace Top 100 Special Report 2012*. PwC and Flight International. Retrieved 20 March 2013 from pdf.pwc.co.uk/aerospace-top-100-special-report-2012.pdf

63 A|D|S (2013). *Growth Period for Space*. A|D|S. Retrieved 3 April 2013 from www.adsgroup. org.uk/pages/81937807.asp

64 Nichols, S. (2013). Space grabs the higher ground, *Advance*, March–May, p. 29–34.

65 UK Centre for Materials Education (2013). *What is Materials Science and Materials Engineering*. The Higher Education Academy. Retrieved 20 March 2013 from www. materials.ac.uk/about/whatis.asp

66 IOM3 (2013). *What Is Materials Science and Engineering?* The Institute of Materials, Minerals and Mining. Retrieved 20 March 2013 from www.iom3.org/content/what-materials-science-and-engineering

67 Materials UK (2008). *Materials Education and Skills: A Wake Up Call*. Materials UK. Retrieved 20 March 2013 from www.matuk.co.uk/docs/MATUK-EduReport(Web).pdf

68 www.reinforcedplastics.com/view/29053/startlink-composite-house-built-in-uk/

69 E-MRS and ESF (2011). *Materials for Key Enabling Technologies*. European Materials Research Society and European Science Foundation. Retrieved 21 March 2013 from www. gfww.de/pdf12/Materials_for_KETs_incl_App2_3.pdf

70 Foresight (2010). *Technology and Innovation Futures: UK Growth Opportunities for the 2020s*, Foresight Horizon Scanning Centre, Government Office for Science. Retrieved 21 March 2013 from www.bis.gov.uk/assets/foresight/docs/general-publications/10-1252-technology-and-innovation-futures.pdf

71 King, D., Inderwildi, O., and Carey, C. (2009). Advanced aerospace materials: past, present and future, *Aviation and the Environment*, pp. 22–7.

72 Bogue, R. (2012). Smart materials: a review of recent developments, *Assembly Automation*, 32(1), pp. 3–7.

73 Lantada, A., and Morgado, P. (2012). Rapid prototyping for biomedical engineering: current capabilities and challenges, *Annual Review of Biomedical Engineering*, 14, pp. 73–96.

74 Papaspyrides, C., Pavlidou, S., and Vouyiouka, S. (2009). Development of advanced textile materials: natural fibre composites, anti-microbial, and flame-retardant fabrics. *Proceedings of the Institution of Mechanical Engineers, Part L: Journal of Materials: Design and Applications*, 223(2), pp. 91–102.

75 BIS (2011). *Innovation and Research Strategy for Growth*. London: Department for Business, Innovation and Skills. Retrieved 21 March 2013 from www.official-documents. gov.uk/document/cm82/8239/8239.pdf

76 www.solaveil.co.uk

77 www.nanocotechnologies.com

78 www.nanoporetech.com

79 Applications of nanotechnology (2012). *Chemical Business*, 26(3), pp. 37–43.

80 www.prospects.ac.uk/materials_engineer_salary.htm. Retrieved 21 March 2013.

81 SEE (2013). *About The SEE*. The Society of Environmental Engineers. Retrieved 23 March 2013 from http://environmental.org.uk/index.php?page=about-the-see

82 www.mastersportal.eu. Retrieved 23 March 2013.

83 ECUK and RAEng (2007). *Statement of Ethical Principles*. The Royal Academy of Engineers. Retrieved 23 March 2013 from www.raeng.org.uk/news/publications/list/ reports/Statement_of_Ethical_Principles.pdf

84 WFEO (1986). *Code of Environmental Ethics for Engineers*. World Federation of Engineering Organizations. Buenos Aires.

85 http://amec-ukenvironment.com/downloads/pp_419.pdf. Retrieved 21 March 2013.

86 www.idgo.ac.uk. Retrieved 21 March 2013.

87 Davis, B. (2012). Structured thinking, *Environmental Engineering*, 25(1), pp. 27–8.

88 www.strath.ac.uk/civeng/news/smartpaintcouldrevolutionisestructuralsafety. Retrieved 21 March 2013.

89 Renner, M. (2006). *Introduction to the Concepts of Environmental Security and Environmental Conflict*. Institute for Environmental Security. Retrieved 21 March 2013 from www.envirosecurity.org/ges/inventory/IESPP_I-C_Introduction.pdf

90 UNESCO (2012). *World Water Development Report 4*. UNESCO World Water Assessment Programme.

91 www.prospects.ac.uk/environmental_consultant_salary.htm. Retrieved 18 March 2013.

92 Walker, J.M. (1996). *Handbook Of Manufacturing Engineering*, New York: Marcel Dekker.

93 EEF (2012). *UK Manufacturing 2012: The Facts*. London: EEF. Retrieved 2 March 2013 from www.eef.org.uk/NR/rdonlyres/1EC00841-E738-44A3-A837-DFBEC00DF99B/20520/EEF _FactCard_2012.pdf

94 HM Treasury and BIS (2011). *The Plan for Growth*. London: HM Treasury and Department for Business Innovation and Skills. Retrieved 2 March 2013 from www.gov.uk/ government/uploads/system/uploads/attachment_data/file/221514/2011budget_ growth.pdf

95 www.prospects.ac.uk/manufacturing_engineer_salary.htm. Retrieved 1 March 2013.

96 Elliott, C., and Deasley, P. (2007). Creating Systems that Work: Principles of Engineering Systems for the 21st Century. London: The Royal Academy of Engineering. Retrieved

20 November 2012 from www.raeng.org.uk/news/publications/list/reports/Creating_ Systems_that_work.pdf

97 INCOSE UK Ltd (2013). Systems Thinking: Why are Systems Important. Ilminster: UK Chapter International Council on Systems Engineering Ltd. Retrieved 1 March 2013 from www.incoseonline.org.uk/Normal_Files/WhatIs/Systems_Thinking.aspx?CatID=What_ Is_SE

98 INCOSE UK Ltd (2009). What is Systems Engineering? Ilminster: UK Chapter International Council on Systems Engineering Ltd. Retrieved 1 March 2013 from www. incoseonline.org.uk/Documents/zGuides/Z1_What_is_SE.pdf

99 www.baesystems.com

100 www.bmtdsl.co.uk

101 www.thalesgroup.com

102 National HE STEM Programme (2012). Unemployment of Engineering Graduates: The Key Issues. Birmingham: National HE STEM Programme.

103 Mendez, R., and Rona, A. (2010). The relationship between industrial placements and final degree results: A study of engineering placement students, Learning and Teaching in Higher Education, 4(2), pp. 46–61.

104 Brown, G., and Ahmed, Y. (2009). The value of work placements, Enhancing the Learner Experience in Higher Education, 1(1), pp. 19–29.

105 Purcell, K., Elias, P., Atfield, G., Behle, H., Ellison, R., Luchinskaya, D., Snape, J., Conaghan, L., and Tzanakoul, C. (2012). Futuretrack Stage 4: Transitions into Employment, Further Study and Other Outcomes. Higher Education Careers Services Unit.

106 High Fliers Research Limited (2013). The Graduate Market in 2013. London: High Fliers Research Limited. Retrieved 14 January 2013 from www.highfliers.co.uk/download/ GMReport13.pdf

107 E4E (2011). Sandwich Courses in Higher Education: A Report on Current Provision and Analysis of Barriers to Increasing Participation. London: Education for Engineering. Retrieved 14 June 2013 from www.educationforengineering.org.uk/reports/pdf/ Sandwich_course_report.pdf

108 Smith, N., and Monk, M. (2005). An evaluation of the impact of a Year in Industry scheme, European Journal of Engineering Education, 30(2), pp. 181–90.

109 CBI and NUS (2011). Working Towards Your Future. London: CBI and National Union of Students. Retrieved 14 March 2013 from www.nus.org.uk/Global/CBI_NUS_ Employability%20report_May%202011.pdf

110 QAA (2009). Personal Development Planning: Guidance for Institutional Policy and Practice in Higher Education. 2nd ed. Cheltenham: Quality Assurance Agency.

111 Reissner-Roubicek, S. (2012). 'The guys would like to have a lady': The co-construction of gender and professional identity in interviews between employers and female engineering students, Pragmatics, 22, pp. 231–54.

Endnotes

112   IES (2013). What is [a] Chartered environmentalist. London: The Institution of
        Environmental Sciences. Retrieved 16 March 2013 from www.ies-uk.org.uk/chartered_
        environmentalist

113   Vitae (2013). What Do Researchers Do? Early Career Progression of Doctoral Graduates.
        London: The Careers Research and Advisory Centre Ltd. Retrieved 16 March 2013 from
        http://www.vitae.ac.uk/CMS/files/upload/What-do-researchers-do-Early-career-
        progression-2013.pdf

114   Harrison, M. (2012). Jobs and Growth: The Importance of Engineering Skills to the UK
        Economy. London: The Royal Academy of Engineering. Retrieved 20 November 2012
        from www.raeng.org.uk/news/publications/list/reports/Jobs_and_Growth.pdf

115   www.civilservice.gov.uk

116   www.iop.org/education/teach/itts/page_52632.html

117   http://ec.europa.eu/environment/chemicals/nanotech/#definition

118   http://investinganswers.com

Further resources and information

UK engineering professional and regulatory bodies

Chartered Institution of Wastes Management – www.ciwm.co.uk

Chartered Institution of Water and Environmental Management – www.ciwem.org

Energy Institute – www.energyinst.org

Engineering Council UK – www.engc.org.uk

Institute of Ecology and Environmental Management – www.ieem.net

Institute of Environmental Management and Assessment – www.iema.net

Institute of Highway Engineers – www.theihe.org

Institute of Materials, Minerals and Mining (IOM3) – www.iom3.org

Institute of Measurement and Control – www.instmc.org.uk

Institute of Operations Management – www.iomnet.org.uk

Institute of Physics (IOP) – www.iop.org/education/teach

Institute of Water – www.instituteofwater.org.uk

Institution of Chemical Engineers – www.ichem.org

Royal Aeronautical Society – www.aerosociety.com

Royal Institution of Chartered Surveyors – www.rics.org

Society for the Environment – www.socenv.org.uk

The Chartered Institution of Building Services Engineers – www.cibse.org

The Chartered Institution of Highways and Transportation – www.iht.org

The Institution of Civil Engineers – www.ice.org.uk

The Institution of Engineering and Technology – www.theiet.org

The Institution of Environmental Sciences – www.ies-uk.org.uk

The Institution of Mechanical Engineers – www.imeche.org

The Institution of Structural Engineers – www.istructe.org.uk

The Society of Environmental Engineers – www.environmental.org.uk

International engineering professional and regulatory bodies

American Institute of Aeronautics and Astronautics, Inc. – www.aiaa.org

ASEAN Federation of Engineering Organisations – afeo.org

Commonwealth Engineers' Council – www.ice.org.uk/About-ICE/What-we-do/Commonwealth-Engineers-Council

European Federation of National Engineering Associations – www.feani.org/site

European Network for Accreditation of Engineering Education – www.enaee.eu

Federation of Engineering Institutions of Southeast Asia and the Pacific – www.feiap.org

International Engineering Alliance – www.ieagreements.org

World Federation of Engineering Organisations – www.wfeo.net

Sector organisations and reports

+Plastic Electronics – www.plusplasticelectronics.com

ADS Group Ltd – www.adsgroup.org.uk

British Dam Society – www.britishdams.org

British Geotechnical Association – www.britishgeotech.org.uk

British Hydrological Society – www.hydrology.org.uk

British Tunnelling Society – www.britishtunnelling.org.uk

Central Dredging Association – www.dredging.org

Confederation of British Industry (CBI) – www.cbi.org.uk

Construction Industry Council – www.cic.org.uk

Education for Engineering – www.educationforengineering.org.uk

EEF: The Manufacturers' Organisation – www.eef.org.uk

Further resources and information

Electronics, Sensors, Photonics Knowledge Transfer Network – https://connect.innovateuk.org/web/espktn

European Materials Research Society (E-MRS) – www.emrs-strasbourg.com

International Commission on Irrigation and Drainage – www.icid.org.uk

International Council on Systems Engineering (INCOSE UK) – www.incoseonline.org.uk

Materials UK – www.matuk.co.uk

Offshore Engineering Society – www.oes.org.uk

PwC Aerospace Top 100 Special Report 2012 – www.pwc.co.uk

Railway Civil Engineers' Association – www.rcea.org.uk

RedR – www.redr.org.uk

Royal Academy of Engineering – www.raeng.org.uk

Semta: Skills Council for Science, Engineering and Manufacturing Technologies – www.semta.org.uk

Society for Earthquake and Civil Engineering Dynamics – www.seced.org.uk

The Engineering the Future Alliance – www.engineeringthefuture.co.uk

The Global Society for Contamination Control – www.gsfcc.org

The Society of Motor Manufacturers and Traders Ltd – www.smmt.co.uk

Tomorrow's Engineers – www.tomorrowsengineers.org.uk

Transport Planning Society – www.tps.org.uk

UK Centre for Materials Education – www.materials.ac.uk

UK Plastic Electronics – www.ukplasticelectronics.com

WaterAid – www.wateraid.org/uk

Wind Engineering Society – www.windengineering.org.uk

Governmental departments and organisations

Defence Engineering and Science Group – www.gov.uk/defence-engineering-and-science-group

Department for Business, Innovation and Skills – www.bis.gov.uk

Department for Transport – www.dft.gov.uk

Environment Agency – www.environment-agency.gov.uk

Health and Safety Executive – www.hse.gov.uk

Highways Agency – www.highways.gov.uk

Office of the Qualifications and Examinations Regulator (Ofqual) – www.ofqual.gov.uk

Technology Strategy Board – www.innovateuk.org

UNESCO – www.unesco.org

Non-engineering professional and regulatory bodies

Association of Chartered Certified Accountants (ACCA) – www.accaglobal.com

Association of International Accountants (AIA) – www.aiaworldwide.com

Audit Commission – www.audit-commission.gov.uk

Bank of England – www.bankofenglandjobs.co.uk

Chartered Institute of Journalists – www.cioj.co.uk

Chartered Institute of Management Accountants (CIMA) – www.cimaglobal.com

Chartered Institute of Public Finance and Accountancy (CIPFA) – www.cipfa.org.uk

Chartered Management Institute – www.managers.org.uk

Institute and Faculty of Actuaries (IFoA) – www.actuaries.org.uk

Institute for Environmental Security – www.envirosecurity.org

Institute of Career Guidance (ICG) – www.icg-uk.org

Institute of Chartered Accountants in England and Wales (ICAEW) – www.icaew.com

Institute of Financial Accountants (IFA) – www.ifa.org.uk

Institute of Financial Services – www.ifslearning.ac.uk

Institute of Leadership and Management – www.i-l-m.com

The Chartered Insurance Institute (CII) – www.cii.co.uk

Resources on engineering

Association of Aerospace Universities – www.aau.ac.uk

Chemical Engineering in Action – www.chemicalengineering.org

Electronics Weekly – www.electronicsweekly.com

Engineer Girl – www.engineergirl.org

Engineering Development Trust (EDT) – www.etrust.org.uk

Engineering Development Trust: First Edition – www.etrust.org.uk/first_edition. cfm

engineeringGateways – www.engineeringgateways.co.uk

EngineeringUK – www.engineeringuk.com

Engineers Without Borders (EWB) – www.ewb-uk.org

Evecracker – www.evecracker.com

Every Woman – www.everywoman.com

Facebook – www.facebook.com

Future Engineering Talent – www.futureengineeringtalent.co.uk

LinkedIn – www.linkedin.com

Lloyd's – www.lloyds.com

NASA Virtual Skies – virtualskies.arc.nasa.gov

Science and Technology Facilities Council – www.scitech.ac.uk

Scottish Credit and Qualification Framework – www.scqf.org.uk

The Chemical Engineer Today Magazine – www.tcetoday.com

The Diploma in Engineering – www.engineeringdiploma.com

Why Not Chem Eng – www.whynotchemeng.com

WISE – www.wisecampaign.org.uk

Women in Engineering – womeninengineering.org.uk

Women's Engineering Society – www.wes.org.uk

Career support

Adzuna – www.adzuna.co.uk

Apprenticeships in Scotland – www.apprenticeshipsinscotland.com

Army Careers – www.army.mod.uk

Association of Engineering Doctorates – www.aengd.org.uk

Association of Graduate Careers Advisory Services (AGCAS) – www.agcas.org.uk

Association of Graduate Recruiters – www.agr.org.uk

Careers in Aerospace – www.careersinaerospace.com

Civil Service – www.civilservice.gov.uk

Creative Skillset (Sector Skills Council for creative media) – www.creativeskillset.org

Department for Education, Get into Teaching – www.education.gov.uk/get-into-teaching

Financial Skills Partnership (FSP) – www.financialskillspartnership.org.uk

FindAMasters – www.findamasters.com

FindAPhD – www.findaphd.com

Go Wales – www.gowales.co.uk

Gradcracker – Careers for Science, Engineering and Technology – www.gradcracker.com

Graduate Acceleration Programme in Northern Ireland (GAP) – www.gapni.com

High Fliers Research Limited – www.highfliers.co.uk

Higher Education Careers Services Unit (HECSU) – www.hecsu.ac.uk

Jobs.ac.uk – www.jobs.ac.uk

Local Government Jobs – www.lgjobs.com

National Apprenticeship Service – www.apprenticeships.org.uk

National Audit Office – www.nao.org.uk

National Council for the Training of Journalists (NCTJ) – www.nctj.com

National HE STEM Programme – www.hestem.ac.uk

National Women's Network – www.national-womens-network.co.uk

New Route PhD (NewRoutePhD™) – www.newroutephd.ac.uk

notgoingtouni.co.uk – www.notgoingtouni.co.uk

Periodical Publishers Association – www.ppa.co.uk

Postgrad Solutions – www.postgrad.com

Prospects – www.prospects.ac.uk

Further resources and information

RAF Careers – www.raf.mod.uk/careers

Royal Marines Careers – www.royalnavy.mod.uk/The-Fleet/The-Royal-Marines

Royal Navy Careers – www.royalnavy.mod.uk

Talent Scotland – www.talentscotland.com

Targetjobs – www.targetjobs.co.uk

Teach First – www.teachfirst.org.uk

The Actuary – www.the-actuary.org.uk

The Building Societies Association – www.bsa.org.uk

The Chartered Institute of Loss Adjusters – www.cila.co.uk

The Graduate Talent Pool – graduatetalentpool.direct.gov.uk

The Institute of Mathematics and its Applications – www.mathscareers.org.uk

The Periodicals Training Council – staging.ppa.co.uk/jobs-careers-and-training/the-periodicals-training-council-ptc

UK Postgraduate Application and Statistical Service – www.ukpass.ac.uk

University Jobs – www.jobs.ac.uk

Welsh Government (apprenticeship service) – www.wales.gov.uk/apprenticeships

Miscellaneous

Coursera – www.coursera.org

Edexcel – www.edexcel.com

Erasmus – www.erasmusprogramme.com

Khan Academy – www.khanacademy.org

OCR – www.ocr.org.uk

Further reading

Blockley, D. (2012). *Engineering: A Very Short Introduction*. Oxford: OUP.

Browne (2011). *For the Engineering Leaders of Tomorrow: Two Lectures by Lord Browne of Madingley. President, The Royal Academy of Engineering 2006–2011.* London: The Royal Academy of Engineering.

Elliott, C., and Deasley, P. (2007). *Creating Systems that Work: Principles of Engineering Systems for the 21st Century*. London: The Royal Academy of Engineering.

Engineering the Future (2011). *Infrastructure, Engineering and Climate Change Adaptation: Ensuring Services in an Uncertain Future*. London: The Royal Academy of Engineering.

Harrison, M. (2012). *Jobs and Growth: the Importance of Engineering Skills to the UK Economy: Royal Academy of Engineering Econometrics of Engineering Skills Project, Final Report*. London: The Royal Academy of Engineering.

King, D. (2010). *Engineering a Low Carbon Built Environment: The Discipline of Building Engineering Physics*. London: The Royal Academy of Engineering.

McCarthy, M. (2009). *Engineering: A Beginner's Guide*. Oxford: Oneworld Publications.

Muir-Wood, D. (2012). *Civil Engineering: A Very Short Introduction*. Oxford: OUP.

Advertiser index